Lecture Notes in Economics and Mathematical Systems

Managing Editors: M. Beckmann and W. Krelle

323

M. B. M. de Koster

Capacity Oriented Analysis and Design of Production Systems

Springer-Verlag
Berlin Heidelberg New York London Paris Tokyo

Author

M. B. M. de Koster
Consultants and Engineers Groenewout B. V.
P.O. Box 3290
NL-4800 DG Breda, The Netherlands

This research was supported by the Netherlands Organization for Scientific Research, NWO.

ISBN 3-540-50692-6 Springer-Verlag Berlin Heidelberg New York
ISBN 0-387-50692-6 Springer-Verlag New York Berlin Heidelberg

Printing and binding: Druckhaus Beltz, Hemsbach/Bergstr.
2142/3140-543210

To Elly

Acknowledgements

The author is much indebted to prof. dr. J. Wijngaard, under whose inspiring supervision this book has been prepared at Eindhoven University of Technology. The author also expresses his gratitude to professors J. Wessels, W.H.M. Zijm and J.W.M. Bertrand for giving advise and criticism to earlier versions of the manuscript. Some stimulating discussions with dr. J. Praagman led to an improvement of section 2.6. Mr. A. Smith–Hardy corrected the use of English in the text. Finally, the author wishes to thank Jolanda Nelissen for typing an earlier version of the text.

PREFACE

In production systems there are often capacity oriented performance objectives, like a desired total throughput, a desired average throughput time and average work–in–process. Such performance objectives are expressed in "units of products" rather than in specific product types.

This book presents a way of modeling and analyzing production systems so, that such capacity oriented performance criteria can be measured in a simple way. The model consists of three basic elements.

1. The product types in the system are aggregated.
2. The product flow is modeled as being continuous.
3. The machines in the model have a finite number of states. Each state has a phase–type sojourn distribution and an associated production speed. Transitions between the states are determined by an irreducible Markov transition matrix.

In the book both the mathematical properties and the practical applicabilities of the model are investigated.

The model is extensively analyzed for various layouts, like flow lines, assembly–disassembly systems and networks where parallel machines share common buffers. Furthermore various ways of controlling the product flow in the model are investigated, such as Base Stock Control, Workload Control, control by finite buffers and control by the Reorder Point System. An approximation technique is developed for a quick estimation of performance measures like throughput and average work–in–process, for networks with layouts and control techniques like those above–mentioned.

By means of three multistage, multiproduct case situations, it is demonstrated that the model is easily applicable. Furthermore, by comparing the model with detailed simulations of the cases, it is shown that the approximation technique gives a good prediction of the capacity performance of the systems.

This book is related to two main research streams on production systems. The first stream is the research on performance analysis of single–product production systems with finite buffers. The main tools in this area are queueing–network models and techniques. The second research stream deals with problems of control structures in production systems. This book is a first attempt to bring together these two streams, by providing modeling and analysis techniques.

TABLE OF CONTENTS

CHAPTER 1

AGGREGATE CONTROL OF PRODUCTION SYSTEMS

1.1 INTRODUCTION

In the design of production systems one is often interested in capacity oriented, rather than detailed performance objectives. An example of a detailed performance objective is the timely delivery of specific product types. Capacity oriented objectives, also called "aggregate" objectives, are objectives with respect to the global behavior of the system, expressed in some aggregate product unit. Examples of such capacity oriented objectives are the realization of a certain production volume (expressed as "products" per year), or the achievement of short average lead times and little work–in–process. It is the task of production control to realize both a good detailed and a good capacity performance. Consider the case of a manufacturer of coin validators (a coin validator is the tool that evaluates coins in coffee machines for instance). This manufacturer is designing a new production line. He has a good idea about what the market will be: worldwide he can sell about 550,000 coin validators a year. The production line has to be able to manufacture this amount of products. The manufacturer does not know in advance to what countries he will sell nor in what quantities. This creates a problem, because different countries demand different coin validators, with different manufacturing characteristics. He has to make sure that, whatever the eventual required product mix will be, he can fulfill this demand (at reasonable costs). The objective with which the engineer is faced, is to design a production system that is able to realize the aggregate target: a total production volume of at least 550,000 items, flexible enough to produce all possible mixes over country variants.

From this case, which is treated in Chapter 7, it will be clear that in design situations the detailed production characteristics, like production volumes per product, are often unknown. The only things that *are* known to some extent are

capacity oriented objectives: a certain desired aggregate throughput, average aggregate throughput time, aggregate service degree, aggregate work–in–process, etc. In existing production systems capacity oriented objectives also play a role.
Consider the case of a truck manufacturer who faces a constant yearly demand for about 34,000 trucks. Within a few years there are no big changes in this number. Although he knows the total demand, he does not know what specific trucks will be demanded, and since he is in principle able to manufacture a few million different truck types, all with different production characteristics, he wonders what the consequences of this volume and this variety will be to his production system. He also has other problems: competition is strong in his branch and to increase his delivery performance he wants to reduce lead times and work– in–process. To what extent is this possible? These are again questions with respect to aggregate objectives.

In an attempt to model production situations as those above–mentioned, it is evident that modeling all individual products is not appropriate. In the first place, the eventual production levels of the individual products are not known and in the second place, it is unnecessarily cumbersome. An aggregate (or capacity) model, where the products are aggregated ("coin validators" in the first example, "trucks" in the second example), will be sufficient to describe the situation and to evaluate its aggregate performance. In Chapter 7, it is shown for these examples that it is possible indeed to model the production systems at capacity level. In such aggregate models a certain production control has to be applied. Production control in an aggregate model is in fact "aggregate control", because the (aggregate) production volume, lead times of aggregates, service degree of aggregates and aggregate work–in–process are controlled. The aggregate control in the model can also be applied in the real system (see below).

This modeling and analysis of production systems at capacity level and the evaluation of aggregate control rules (also called *capacity control*) is the subject of this text. Note, that in single–product production situations the detailed and capacity level of describing the production system, coincide. The importance of modeling systems at capacity level is in the first place due to the fact that objectives are expressed in aggregates, which leads to aggregate models. Controlling the flow in aggregate models leads to aggregate control. There is another reason why it is important to study aggregate models and aggregate production control. It is because many production situations are so complex, that production control has deliberately been split up into different hierarchical levels (at least two). At the higher control level aggregate targets are set and controlled. The lower control level (*item control*) has to make sure that individual items are produced in time. A well–known goods flow control system like Manufacturing Resources Planning (MRP II, see for instance Plossl and Welch, 1979), is based on this idea. Two main levels are distinguished, namely Production Planning (PP) and Material Requirements Planning (MRP I). The PP level is concerned with setting aggregate targets: total production volumes, total budgets for departments, aggregate stock norms and norms for lead times are determined. The MRP I level has to make sure that specific items are produced in time, given the restrictions of PP. For instance, order releases are offset, by means of average lead times determined at PP level.

To determine realistic aggregate objectives and realistic norms at the capacity control level in two–level production control, it is necessary to have available aggregate models, a control structure for capacity control and a method for analyzing such models.

Splitting up production control into a capacity level and an item level has been advocated by several authors. Bemelmans (1986) has shown that, for certain single–stage multiproduct production/inventory systems, it is close to optimal to let total production depend on total inventory only, and to let the allocation of production only depend on the individual inventories. Wijngaard (1984) has argued that the decomposition of capacity and item control can be extended to multistage production systems. For related literature see also Hax and Meal (1975), Bitran and Hax (1977), Bitran et al. (1981,1982), Van Beek (1981), Hax and Candea (1984), Meal (1984), Bertrand and Wijngaard (1986) and Kimemia and Gershwin (1983).

As an example of aggregate objectives and capacity–control decisions, we consider the case of a differential–gear factory. The case is treated in more detail in Chapter 7. Production control in this factory will appear to be rather complicated, and therefore a separation into two control levels seems natural.

The differential–gear factory supplies differential–gear parts to the truck–assembly factory referred to above. These gear parts are manufactured from raw materials (iron casts) by cutting, drilling and grinding operations and by heat treatment. About 86% of the total gear–part production consists of cogwheel/bevel–

gear sets. The number of such final product set types is about 65. The main product flow in the cogwheel/bevel–gear manufacturing is sketched in Fig. 1.1.1. Triangles denote stock points, rectangles work centers.

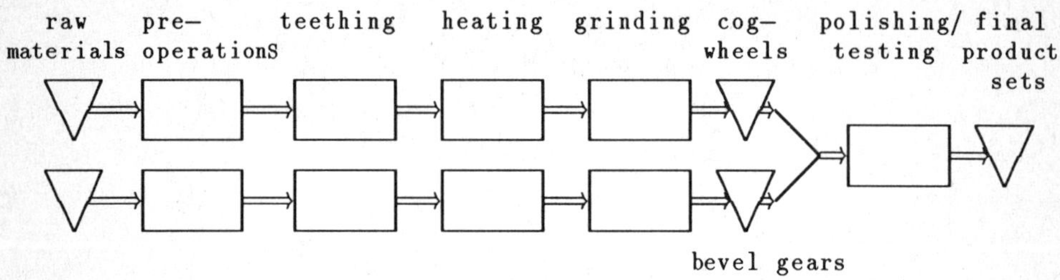

Figure 1.1.1. The main flows in the differential gear factory

The 65 different cogwheel/bevel–gear sets can be divided into four main product types (A, B, C and D). In the preoperations capacity group all cogwheels and bevel gears have to pass through three different operations sequentially (mainly cutting and drilling). All cogwheel types are sequentially processed by the same three machines. The bevel–gear types pass through the first two operations on the same two machines, however the last operation is performed by two dedicated machines (one for types A and D and one for B and C). There are relatively long set–up times for most products, which makes lot sizing necessary in order to achieve the required production efficiency. A skilled operator is needed for each machine operation. In the teething capacity group, the cogwheels and bevel gears are teethed by several parallel teething machines. Furthermore the metal parts are deburred here. Subsequently, the gear parts require two oven operations. Lot sizing is necessary in one of the ovens for the cogwheels and in both ovens for the bevel gears, because of special oven conditions for different parts, to say nothing of transportation restrictions. After heat treatment the parts are ground. Then sets of (corresponding) cogwheels and bevel gears are formed from synchronization stocks prior to polishing (see Fig 1.1.1), then polished and tested, after which they are transported to the assembly factory.

The production planning in the gear factory is done on the basis of the assembly plan for the trucks in which the gear sets are assembled. Via a rolling schedule, the assembly plan is known over a horizon of four weeks. The number of trucks per day in the plan is about constant. The result of the explosion is a time–phased demand for cogwheel/bevel–gear sets.

What makes production control difficult in the factory is the presence of many sources of capacity fluctuations:

- Variation in service times. Different items of the same type may require different service times at the same machine (for instance, because an operator has to control several machines simultaneously).
- Different products have different capacity requirement patterns at the capacities. For instance, in the preoperations the product types have different capacity requirements for the two major capacities. Three types (A, B and D) have faster production rates at the first capacity than at the second. Type C has faster rate at the second capacity than at the first. Simultaneous production of these types causes capacity fluctuations.
- Several machines are subject to failures.
- Uncertainty in the availability of operators. Operators may become ill.
- Capacity fluctuations because of set–ups for different products.

One way to cope with these capacity fluctuations is by putting some stock into the system, in order to ensure at least sufficient capacity utilization (of men and machines). The production control in the factory is then split up into two hierarchical levels. The higher level of control (*capacity control*) has to guarantee a good aggregate performance: short average lead times and a satisfactory capacity utilization. This can be realized by controlling the aggregate inventories and work–in–process.

The lower level of control (*item control*) has to make sure that production runs for the right items are started at each release point. The items are produced according to urgency. It is possible to decide which product type has to be produced, only a few days before the date that it is due. This is possible, because the total processing time for a single batch is not more than one and a half days. Also, if some waiting time is included because another batch may still be in production, a priority batch can still be produced within two days.

Questions that have to be answered at capacity level in this factory are, for instance:

- the yearly demand for trucks implies that the yearly number of differential gears has to be about 34,400. However, the daily product mix is not known. Is it possible to reduce the throughput times, while simultaneously realizing the desired throughput for all possible product mixes?

- Can the stocks and work–in–process be reduced?
- When are products to be released in the factory and how has the stock in the system to be controlled?

In this text, tools will be developed for the analysis of production systems at an aggregate level and the evaluation of control policies. Such tools can be used to answer questions such as those occurring in the differential–gear factory. For the case of the differential–gear factory they will be considered in Chapter 7.

1.2 PROBLEM FORMULATION

The problems arising in the control of the differential–gear factory, i.e. the determination of capacity control rules as well as item control rules and the determination of norms for aggregate performance, can be encountered in many practical situations. This text focuses on *capacity analysis* and *capacity control*, (the latter is also described as *capacity coordination* or *aggregate production control*) in multistage, multiproduct production systems. The emphasis will mainly be on flow structures: all products have about the same routing over the capacities. Yet other structures will also be discussed: groups of parallel machines, assembly– disassembly structures. Job–shop–like structures however are left out of scope. This text is intended to provide tools (models, approximation methods) for the capacity analysis of such flow–oriented production systems. Those tools can also be used for the design of capacity control rules and for obtaining answers to questions about reduction of throughput times, increase of thoughput, reduction of work–in–process. In general, these tools have to be used for

- Determination of aggregate performance and norms for performance, in particular for throughput, throughput times and average work–in–process. These norms are not independent of each other: e.g. there is a linear relation between average throughput time and average work–in–process.
- Some kind of aggregate production control has to be applied. It is important to determine *how* performance norms should be kept at norm. The *structure* of the control system has to be determined. On the basis of what information must work release decisions be taken?

These two issues are now discussed in more detail.

To determine aggregate performance and interrelationships between aggregate norms, a production–system model will be developed in which the relation between inventories, work–in–process, throughput and throughput time can be quantified, and in which the effect of different control methods can be evaluated. Such a model has to satisfy certain conditions. It must be analyzable and it must be possible to draw general conclusions from it. Furthermore, the results must be easy to apply. The basic properties of the model will be discussed in section 1.4.

As far as the structure of the capacity control rules in the model is concerned, it has to be decided how the production levels of the capacities depend on the state of the system. These production levels will have to depend on the aggregate demand, on the aggregate inventories and on the state of other capacities. Information on (aggregate) inventories between the stages is necessary anyway, since full–speed production at a certain stage is pointless if there is already a lot of inventory immediately downstream of the stage. Capacity control is interpreted here as control of the reaction of capacities to aggregate inventories. Hence demand is only indirectly incorporated via its impact on inventories for final products. In section 1.3 several structurally different ways in which production levels can depend on inventories, are discussed.

Capacity control can be applied at different levels, depending on how the production stages are chosen. In this text, the stages will in the first place be interpreted as simple machines without work–in–process, but later it will be found that a similar analysis can be carried out if the stages consist of small groups of machines (section 2.4), or even self–contained Production Units (PU), in the sense of Bertrand and Wijngaard (1986): groups of machines with work–in–process (Chapter 6).
Only stationary situations are considered: stationary behavior of capacities and demand. Stationarity of demand may be difficult to check, but situations where production is based on a fixed plan (in volume), like in the differential–gear factory fall under the scope of the text.

1.3 CAPACITY CONTROL STRUCTURES

Two different control structures can be distinguished to control the aggregate production of the stages. These control structures are denoted by local and integral control. In local control, discussed in subsection 1.3.1, only the next inventory is used to control the production volume of a certain stage. In integral control, inventories are used in a more complex (integral) way to control production. Integral control structures are discussed in subsection 1.3.2. The Reorder–Point System and Kanban are examples of local control. MRP, the Base–Stock System, Workload Control and OPT (Optimized Production Technology) are examples of integral control. Although most of these control systems are designed for control at item level, here only the distinction between their local and integral nature is stressed.

1.3.1 Local control structures

In a **local control** structure, the production volume of a stage is assumed to be determined by the inventory levels in the immediate upstream and downstream buffers. See Fig 1.3.1.

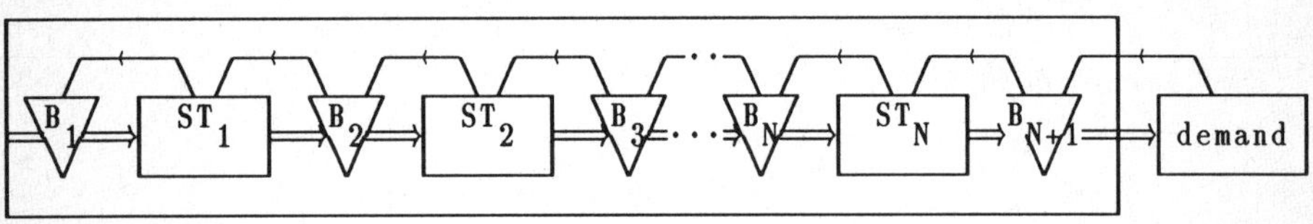

Figure 1.3.1. N–stage locally controlled flow line. ⟵ : information flow. ⟹ : material flow. ST stands for production stage. B stands for buffer.

The system in Fig. 1.3.1 works as follows. The demand for end items enters the system at stockpoint B_{N+1}. The contents of this stockpoint triggers production of ST_N. ST_N in turn, takes its material from stockpoint B_N, which is also locally controlled and hence ST_{N-1} is triggered, etc.

Upstream buffers only influence the production volume if they are empty. Downstream buffers may influence the production volume in various ways. The simplest case is the situation where stages are machines and buffers have finite capacity: each machine tries to fill the downstream buffer. It becomes blocked if the downstream buffer is full. Slightly more complicated is the situation where buffers have several trigger levels. If the buffer content increases, to such a trigger level, then the speed of the upstream stage changes.

This local way of controlling aggregate production is similar to the Reorder–

Point System (RPS, see for instance Hadley and Whitin, 1963 or Silver and Peterson, 1985) for control at item level. RPS is concerned with two questions, namely when products have to be ordered from the production stages and how much. The answer to the first question is to check whether the stock for the particular item has fallen below a so-called *reorder point.* If this is the case, then a certain quantity of products is ordered. The reorder point is denoted by s or B, depending on whether the review is done periodically or continuously. The reorder points are based on the demand during the lead times.

The quantity ordered is usually a fixed batch, denoted by Q, or the batch size is the difference between a fixed order up to level S and the actual level. Combination of these possibilities leads to the four well-known strategies $<s,S>$, $<B,S>$, $<s,Q>$, $<B,Q>$.

Applying RPS has some well-known drawbacks. An important one is the delay of information with respect to the demand, as experienced by stockpoint B_{N+1}, to other stockpoints. If the demand for a certain final product increases, then ST_N may have to adapt its reorder level after some time. Due to the positive lead times the other stockpoints will also adapt their reorder levels with a certain delay. This can lead to instability in the chain (see also Forrester, 1961). As already mentioned, we are mainly interested in stationary situations, so that sudden changes in (average) demand are excluded. Nevertheless, delay of information on the demand rate is also actual in the line in Fig. 1.3.1. Suppose for instance, that ST_N is an unreliable machine which has broken down when demand enters the system. Then there is a time delay before upstream stages adapt their production speeds.

In the Base Stock System (BSS), to be discussed next, the information on demand for final products is not delayed, but is immediately transmitted to all upstream production stages.

1.3.2 Integral control structures

In the text, two different integral control structures are considered for capacity control, namely the Base-Stock System and Workload Control. Applying integral control structures at capacity level seems to have an advantage over local control: the total work-in-process can be better restricted, while preventing important capacities from becoming idle and hence ensuring the throughput.

The Base Stock System. As already mentioned, the demand for aggregate final products in the Base Stock System (BSS) is directly communicated to all upstream stages. See Fig. 1.3.2.

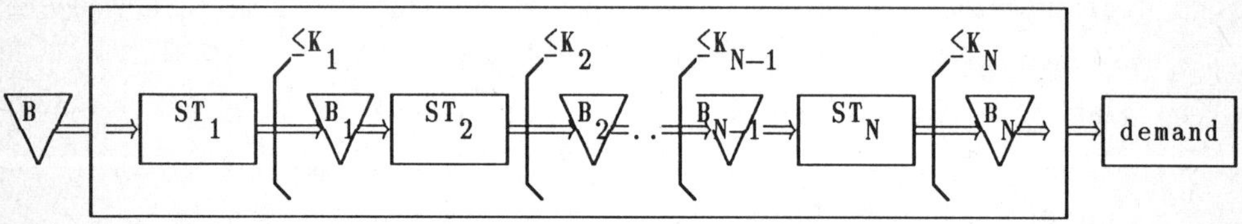

Figure 1.3.2. N–stage flow line controlled with BSS. ⟹: material flow.

In BSS, the demand experienced at stockpoint B_N, is directly communicated to all production stages via echelon buffer contents. A trigger level is defined for the aggregate product for each production stage. The trigger level for ST_i is denoted by K_i. All levels are measured in the same unit. They are chosen so that $K_1 \geq K_2 \geq .. \geq K_N$. As in local control, production of a stage is stopped if the downstream stock increases to the trigger level. However, it is not the stock in the downstream buffer which is compared with this trigger level, but all stock and work–in–process downstream of the stage (the part of the flow downstream of a production stage, including that stage, is called a (right) echelon). If the demand at stockpoint B_N causes the stock in any echelon to drop below its trigger level, then the immediate upstream stage of the echelon tries to replenish it. The aim of BSS is to keep the total work–in–process in the system restricted, while fulfilling demand as much as possible. The delay of information on the demand no longer occurs in BSS.

For a detailed description of BSS in multiproduct situations see Magee (1958), Timmer et al. (1984) and Silver and Peterson (1985). Clark and Scarf (1960) showed that control with BSS is optimal with respect to a certain cost function depending on inventory–holding costs and stock out costs, for the case of a flow line with fixed lead times and no capacity constraints.

Using BSS requires more information processing than using local control (in particular, information about all downstream stages), but nevertheless BSS is not very complicated.

Workload Control. Workload control, also called *Balancing Workload* or *Input–Output Planning*, is a control system, originally developed for job shops (Bertrand and Wortmann, 1981, see also Plossl and Welch (1979), Wiendahl (1987), Bechte (1984) and Bertrand and Wijngaard, 1986), but later also used for flow shops (Paumen, 1985). In classical queueing theory, the utilization of capacities is determined by the amount of work offered, which again leads to a certain throughput time. In workload control this idea is reversed. In order to achieve a certain capacity utilization and throughput time, the workload (expressed in number of batches in the system, or the remaining processing time for all operations not yet processed on certain capacities) is controlled.

The way it is used in this text is illustrated in Fig. 1.3.3. For each stage i there is a norm for the workload in the production system preceding the stage. This norm is denoted by K_{i-1}. Stockpoint B, from which work is released to the system, reacts to the workloads of all downstream stages. Work release to the system is stopped as soon as the workload in any of the left echelons (a left echelon is the part of the production system preceding a production stage) reaches its norm. The aim is, by controlling the input, to prevent idleness of the important capacities and to load them as well as possible, while keeping the work–in–process restricted. The K_i can be chosen such that some of them are redundant, depending on which capacities are critical. K_i is redundant if $K_{i+j}<K_i$, for some j. All K_i are active, if $K_1 \leq K_2 \leq .. \leq K_N$.

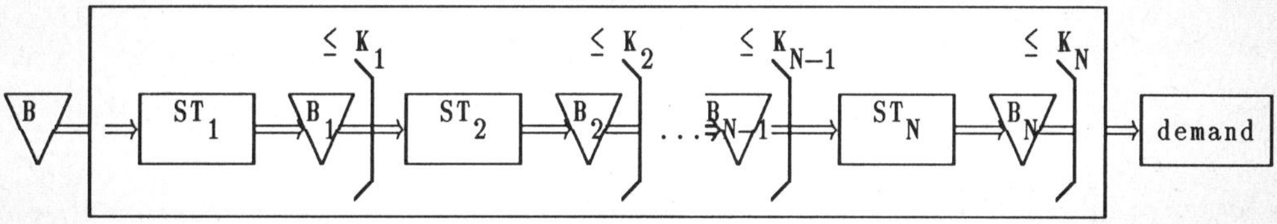

Figure 1.3.3. N–stage flow line controlled with Workload Control. ⟹ : material flow.

The control structures presented in this section are used in this text to control the production level of the capacities. In Chapter 5 they are compared for their performance with respect to throughput and average work–in–process.

1.4 CAPACITY MODELING OF PRODUCTION SYSTEMS

There are several possible ways to determine capacity performance of production systems and to evaluate the effect of different control rules. The most straight-forward method is to build a detailed simulation model of the system, including all product types and capacities and a detailed control rule. By simulating the system sufficiently long, aggregate performance measures can be measured, like total throughput, average total work–in–process and average throughput time. To evaluate the effect of different control rules in such a model, these control rules can be varied. Although simulation is a powerful tool, it has certain drawbacks. First, the simulation models will in general be complicated, with many parameters involved. It is difficult to obtain structural insights in the relations between the different parameters and aggregate performance. Second, simulation is time consuming. In the design phase of production systems, time will often be insufficient to evaluate all possible scenarios.

Therefore in this monograph another approach is proposed. Production systems are modeled in an aggregate way. The aggregate product flow in the model is controlled by one of the control structures, discussed in the previous section.
The model has to to possess certain properties. It must give insight in relations between work–in–process, throughput time and throughput. Furthermore, it must be possible to use it for the evaluation of different control systems (local or integral) and, what is more, it must enable structural conclusions to be derived, for example, whether it is desirable to use local (aggregate) or more global (aggregate) inventory information for the release of products. Last, but not least, the model must be rich enough in parameter choices, to be able to check insensitivities in these choices.

A special class of models will be used. In the choice of the model, three basic aspects are involved.
First, the product flow is chosen continuous. The reason for this is that, at capacity level, production is measured in man or machine hours and a working capacity entity (man, machine) produces one man or machine hour per hour. Hence, the purest way to model capacities is to use a continuous product–flow model. The problem is of course, that this basically continuous character is disturbed by the fact that production is in discrete units or even in complete batches. At an aggregate level this discreteness may be somewhat less important, but still present. On the other hand, it is not easy to use discrete aggregate

models. In discrete–product models, the choice of the aggregate product is not at all straightforward. Think of the situation where one unit of product 1 requires one hour and a unit of product 2 requires two hours. The aggregate product unit can not be a unit of product 1 or of product 2. Or assume that in stage I the lot sizes are different from those in stage II. Another solution would be to model *all* products in a discrete product model. However, this increases the state space and makes the models very complicated or even intractable. Therefore, in this text the choice is made to model the product flow continuously.
The second feature of the model is, that only one "product type" is considered (hours of production or units of product). The reason for this is that we only want to control aggregate performance. It is assumed that aggregate information and aggregate models are sufficient for this purpose. One has to be careful in the choice of the aggregate product unit. The problem is to express hours of executed work by stage 1 in hours of remaining work for the immediate downstream stage 2, without identification of individual products? This problem arises when the capacity requirements for both stages are not the same for all products. The influence of this complication is less severe, if the capacity requirement ratios differ less and if the buffer–content mix is more stable. Whether or not a one–product continuous model is appropriate, depends on these factors. A more thorough discussion follows in Chapter 6.
The third property is that all sources of randomness are incorporated in the machine behavior. The machines in the model have different states. Each state corresponds to a certain constant production speed. The sojourn time in a machine state is assumed to be exponentially distributed and the transitions from one machine state to another are determined by a Markov transition matrix.

The machine model is sketched in Fig. 1.4.1. The model parameters, such as machine speeds, life rates in the different states and the transition matrices, have to be chosen so, that the model describes the system well at capacity level. Choices have to be made how machine cycle times for different products are incorporated in the model. As an example, the different production rates in the model may correspond to production rates of different products in the system. Sojourn times in each state may correspond to batch processing times of a specific product type and some states may correspond to switch overs (by choosing some production rates zero). This type of modeling will appear to be sufficiently general, at least for the cases discussed in section 2.1 and Chapter 7. Furthermore, it enables us to develop a rather simple analysis method, as will be shown in Chapter 2.

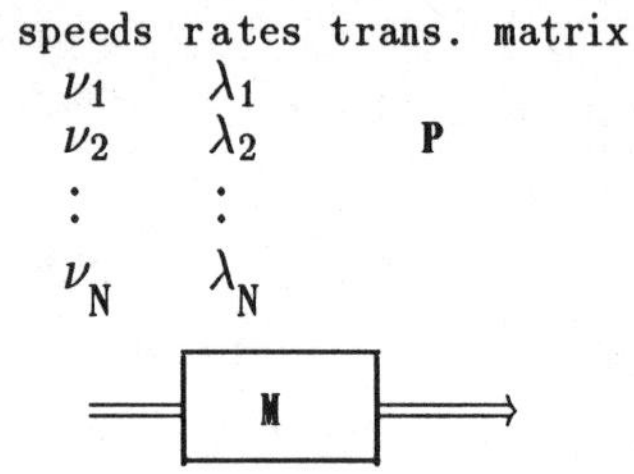

Figure 1.4.1. Machine model used in the text

Summarizing, the three most important features of the model used in this text are:

1. Continuous product flow.
2. Single product type.
3. Capacities with several states and possible machine speeds.

Another factor that justifies the choice for a continuous product–flow model instead of a discrete one, is the fact that continuous models can very well approximate discrete product models (at least as far as the throughput is concerned). This will be demonstrated in section 2.8.
A problem in the use of continuous–flow models is that machines in a such a model cannot contain products. This problem can be solved. Possible solutions are:

- split up the storage and production capacity of machines.
- control the flow by input–output planning. Hence the number of products in the system is constant. If the throughput of the model and the real system are close, then also the throughput time will be correctly modeled, by Little's formula.

It will be shown by three case situations, in section 2.1 and Chapter 7, that the model can be used to describe production systems at capacity level. Aggregate performance of these systems can be predicted fairly well by the performance in the model. The model performance is evaluated by numeric methods, rather than by simulation. Numeric methods have the advantage that they are much faster than simulation of the model, so that they can be used for a quick evaluation of

different scenarios (different control policies, different system parameters).
The results obtained in the text are:

1. The derivation of simple tools for quick numeric evaluation of the (aggregate) performance of production systems. The tools can be used for a quick scan the production system, to give first insights into aggregate performance. They can be used interactively to evaluate different control systems, or to evaluate the effects of buffers in production systems.
2. Guidelines for the choice of the aggregate product. Thses guidelines are provided by the different case situations.

1.5 CONTENTS OF THE TEXT

Although the main objective of this text is to provide tools for the capacity analysis of complicated multistage production systems, we start in Chapter 2 with much simpler systems, consisting of two stages only. Systems such as the differential–gear factory in section 1.1, are too difficult at this stage to start the analysis with. Therefore, section 2.1 describes the simpler case of a two–stage multiproduct production line, namely an insertion department in a consumer electronics factory. The problem is to determine a norm for the total work–in–process, in order to obtain satisfying throughput levels. Modeling of the case as a two–stage continuous product flow line leads to the basic model for the rest of this text: a single product two–stage line with multistate machines and a finite intermediate buffer. Since there is only one buffer, control policies do not really play an important role in this model: the first machine works as long as the buffer is not full. Before treating the case in detail, this two–stage model is analyzed thoroughly in Chapter 2. The model is analyzed with different parameter settings (section 2.3) and with more complicated machines (section 2.4). More complicated triggering policies are investigated in section 2.5. In sections 2.6 and 2.7 the buffer behavior is analyzed more carefully.

After this detailed analysis of the continuous flow model, it is compared with a discrete–product model in section 2.8. Discrete–product models can be approximated by continuous models by approximating the buffer fluctuations. In section 2.9 the case of section 2.1 is analyzed. Norms are established for the total workload in the system.

In Chapter 3, the simple two–stage model is extended to more complex systems, consisting of several machines and buffers. Analytical tools are discussed for the evaluation of such systems. It appears that numerical methods yield difficulties for systems with more than one or two buffers.

Three types of layouts are studied:

1. Flow lines.
2. Assembly–disassembly networks. In these networks, stages may be fed by several buffers (assembly), or stages may supply several buffers (disassembly).
3. Buffer–sharing networks. In these networks, a single buffer may have several upstream or downstream stages.

Chapter 4 deals with approximation techniques for networks with layouts as introduced in Chapter 3. All stages in these networks are locally controlled by finite buffers. An approximation tool is developed (4.1) for the establishment of (approximate) relations between local stock norms and throughput and mean work–in–process. Other buffer–triggering policies are also discussed (in section 4.7).

In Chapter 5 the analysis is extended to non–local control systems. The relationship between stock norms and line performance is investigated for flow lines controlled with the Base–Stock System and with the Workload Control System. The same approximation tool is used as in Chapter 4. Surprisingly, it is found that, for the stationary single–product model there is no real gain in using integral control instead of local control systems. It is shown that for each continuous–flow line, controlled with BSS or WC, it is possible to find a locally controlled line with about the same throughput and total work–in–process. The same is shown to hold for a discrete product model.

In Chapters 2 to 5, the production systems considered were essentially single–product systems. Aggregation problems, stemming from the fact that in reality there are many products, are discussed in Chapter 6. Another problem discussed is, that in multiproduct situations lot sizing may occur. Production situations in which the stages are not simple (groups of) machines, but larger groups of machines with work–in–process (production units) are also considered. It is argued that to a large extent the tools for the analysis of simple stages can also be applied to such production situations.

Finally, in Chapter 7, two cases are described and analyzed by the tools developed in Chapters 4, 5 and 6. The first is a multistage, multiproduct flexible assembly line, where certain electromechanical products are manufactured. For this line, which is controlled by Workload Control, norms are established for the total work–in–process, depending on the annual product mix. The second case describes the differential–gear factory introduced in section 1.1. For this second case, a control rule, based on the amount of work–in–process, is determined so that the production plan can be realized and furthermore the throughput times are reduced.

In short, the contents is sketched in the diagram in Table 1.5.1.

chapter	network layout	control structures	method
2	single product (2.1,2.9:multiproduct) two–stage	local	analytic numeric
3	single product multistage flowlines, ass.–disass,buffersharing	local	analytic numeric
4	single product multistage flowlines, ass.–disass,buffersharing	local	heuristic
5	single product multistage flowlines	integral local	heuristic
6	multiproduct multistage flowlines	integral local	heuristic
7	multiproduct multistage cases	integral local	heuristic simulation

Table 1.5.1. Summary of the contents of the text

This book is related to two main research streams on production systems. The first stream is the research on performance analysis of single–product production systems with finite buffers. Also research on general queueing networks with finite buffers is included. The second research stream deals with problems of control structures in production systems. This book is a first attempt to bring together these two streams, by providing modeling and analysis techniques.

Those readers interested purely in the results, be they technical or applicational, may be best served by sections 1, 7 and 9 in Chapter 2, Chapters 4 (except section 1), 5, 6 and 7. Readers interested in backgrounds and possible extensions of the used models are recommended to read Chapters 2 and 3 as well. The appendices are for readers interested in background information on the models developed in the text. Practitioners with little time, but interested in applicability of the results, are recommended to read the cases in sections 2.1 and 2.9 and in Chapter 7.

CHAPTER 2

CAPACITY ANALYSIS OF TWO-STAGE LINES

In this chapter, a model is developed to describe production systems at capacity level. This model is the one briefly introduced in section 1.4. It is only applied here to simple production systems, consisting of two stages only. It is assumed that these production systems produce a single product type. A multiproduct case situation is considered only in section 2.1. The objective is to determine aggregate performance in such systems: throughput and mean system contents. Via Little's formula, then also the mean throughput time is known. The model will be applied to more complicated multistage, multiproduct production systems in later chapters.

The use of the model is illustrated with a case. Since the factory described in section 1.1 is still too complicated for this purpose, the simpler case of an insertion department is used (section 2.1). This insertion department has a limited number of machines and produces many different products. It is controlled with workload control. The problem in this case is to determine the aggregate throughput of the system as a function of the aggregate workload norm. It is shown that the continuous flow model is fit to describe this situation. At the end of this chapter, in section 2.9, sufficient tools will have been developed to analyze this case.

Before the analysis and mathematical backgrounds of the model are discussed (sections 2.3 till 2.8), first a literature survey of analytical models for two-stage lines with finite intermediate buffers is given in section 2.2.

In section 2.3 we start the analysis with a simple version of the model (two unreliable machines, with constant speeds, exponential life and repair times and a finite intermediate buffer). Several performance measures are calculated for different parameter settings. In section 2.4, a similar model is analyzed, with more speeds per machine. The insertion department can be described by this model.

The performance of the two-stage line is influenced by the exploited buffer control policy. Different buffer control policies are investigated in section 2.5. The buffer behavior for the two–stage line is characterized in section 2.6. The performance of the line is essentially determined by its buffer behavior. There appears to be a high degree of insensitivity in system performance with respect to the exact machine parameters, so long as the mean and variance of increase and decrease of the buffer contents and furthermore the expected lengths of periods of increase and decrease of the buffer contents remain the same. These results are used to approximate complex continuous flow lines by simpler ones (section 2.7). In section 2.8 discrete product two–stage lines and continuous flow lines are compared as to throughput and mean buffer content. For certain cases, a simple relation is given between parameters of discrete and equivalent continuous two–stage lines.

Finally, in section 2.9 we return to the case of section 2.1. A relationship is determined between the workload norm and the throughput for the insertion department, using the tools developed in this chapter (especially section 2.4).

2.1 CASE: AN INSERTION DEPARTMENT

2.1.1 Case description

As an example of the use of continuous flow, single–product–type modeling, consider the case of an insertion department which is part of a consumer electronics factory (see Wijngaard, 1987). In this department, printed–circuit boards (PCBs) are mounted. Subsequently "horizontal" and "vertical" components are inserted mechanically on printed–circuit boards. An electronic component is called "horizontal", if it is two–dimensional (that is, nearly flat) and "vertical", if it is three–dimensional (has a certain height). The insertion of horizontal and vertical components requires different machines. The machines inserting the components on the PCBs are numerically controlled. A sketch of the department with PCB flows and component flows is given in Fig. 2.1.1. There are three horizontal and four vertical inserters, separated by a buffer. Buffers are denoted by triangles in Fig. 2.1.1. The first buffer consists of released PCBs, waiting for insertion. The intermediate buffer consists of PCBs with attached horizontal components, waiting for vertical insertion. A rectangle in Fig. 2.1.1 denotes the combination of an inserter and a so–called "sequencer". A sequencer feeds components to the inserter in the correct insertion sequence. This is important, since the department produces many types of different PCBs (about 80). Each PCB has its specific size and its own number and types of horizontal and vertical components for insertion at predefined locations. The sequencers are fed by a set of large tapes, each containing one type of components. These components are automatically fed into the inserter in the correct sequence, depending on the specific type of PCB being made. Switching between different PCBs requires set-up time, due to replacement of programs of inserters and sequencers, changes in dimension between subsequent PCBs and less often because of replacement of tapes. Therefore production is in variable batches of PCBs.

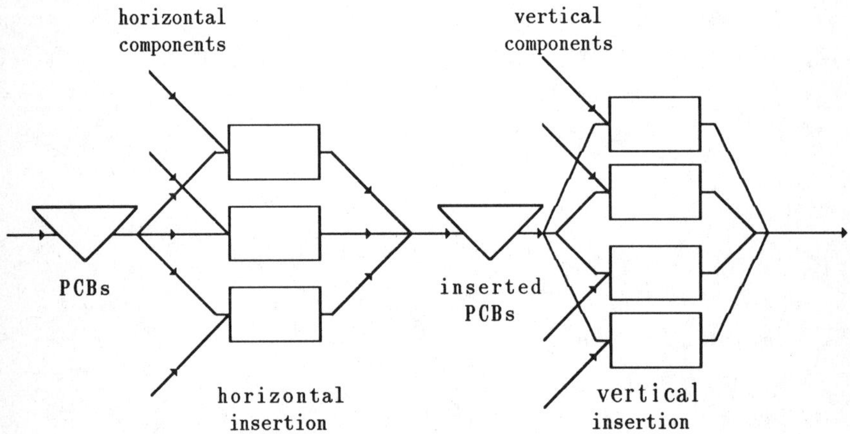

Figure 2.1.1. The insertion department

There are no transportation batches from horizontal insertion to the buffer. Hence the PCBs arrive in the buffer one by one. When necessary, the vertical inserters already start working on the first items of a batch, even before it is completely available. However, in principle, when sufficient batches are available, each vertical inserter operates on its own batch. Machine failures may also occur. In Table 2.1.1 some characteristics of the line are given. The numbers in Table 2.1.1 are averages. The batch sizes (expressed in PCBs), for instance, have a high standard deviation. On the other hand, the machine speeds of the inserters (expressed in components per minute) are almost deterministic and independent of the component type.

The automatic insertion department is only one of the three departments in the factory. In Fig. 2.1.2 the main product flow is indicated. Some side flows of subassemblies which are assembled on the main product are also indicated.

	machine speeds (comp./min)	change–over time (min.)	repair time (min.)	total down fraction	batch size (PCBs)
horiz. ins.	230	3.3	12	10%	105
vert. ins.	90	6.0	11	14%	

Table 2.1.1. Parameters of horizontal and vertical inserters

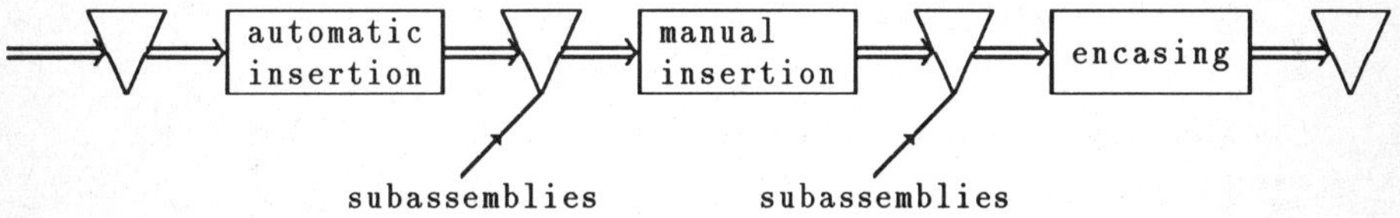

Figure 2.1.2. The goods flow in the production system. In encasing, inserted PCBs are assembled into T.V.s.

The different departments in this factory are coordinated by Goods Flow Control (GFC), which releases orders to the departments. PCBs belonging to a batch are released to the automatic insertion department, if the actual work–in–process (WIP) level (measured in PCBs) within this department is less than the work–in–process norm (denoted by K). If there is too much WIP, then no PCBs are released until the WIP level drops below the norm K. See the description of Workload Control in section 1.3.

At capacity level, the problem is to *determine the work–in–process norm* for this department corresponding to a certain *desired throughput and throughput time.* The throughput time of the insertion line for an individual PCB always consists of horizontal and vertical processing times plus some incidental waiting time if another batch is being processed. Since, on average, a PCB contains about 120 components, it can be seen from Table 2.1.1 that the processing time of a PCB will be short. Since the total number of PCBs in the system is restricted, and PCBs are nearly FIFO (First In First Out) processed by the vertical inserters, also the throughput time of PCBs will be short. It is assumed here that the right bare PCBs and the right components are always available. Hence the only remaining problem is *the dependence of the throughput on the work–in–process norm* K.

In the next subsection a model of the insertion department is proposed.

2.1.2 Modeling the insertion department

In this subsection the automatic insertion department will be modeled. The problem is to determine the relationship between the work-in-process norm of the department and the throughput. The throughput is expressed in finished PCBs per minute. It will be clear that it is impossible to incorporate all 80 different PCBs (and the roughly 8000 different components) in a tractable model. Since the desired throughput is expressed in an aggregate quantity (PCBs per minute) we settle for a single product model. All variables will be expressed in aggregate PCBs.

The problem is now to express the work finished by horizontal insertion (expressed in numbers of PCBs) in work to be done by vertical insertion. Different PCB types require different numbers of horizontal and vertical components. Hence the work content per PCB for vertical insertion differs per PCB type. Yet it is not difficult to solve the problem, because the ratio between horizontal and vertical components is about the same for all PCBs, namely 2:1. An average PCB contains 120 components, 80 of which are horizontal and 40 vertical. The speeds of the horizontal inserters are now expressed in "norm PCBs" per minute, where a "norm PCB" is a PCB containing 80 horizontal components and 40 vertical ones. The speeds of the vertical inserters and the work-in-process can also be expressed in norm PCBs. The aggregate product is now "norm PCBs".

Taking a "norm PCB" as containing 80 horizontal and 40 vertical components, it can be seen from Table 2.1.1 that the processing time of a single PCB is short (in the order of a minute) compared with the production time of a complete batch (in the order of an hour). Therefore it is natural here to model the product flow continuously here with production rates for the machines. Note that a queueing analysis approach with random service times for PCBs at horizontal and vertical inserters is difficult, because the service times of the PCBs at horizontal and vertical inserters are coupled.

It is furthermore assumed that the right components and bare PCBs are available at the inserters at the right times.

Each inserter/sequencer combination is modeled to be in either of the two states: producing at a certain constant speed or down (due to repairs or switch-overs), each with some specific distribution for the sojourn time in that state. The uptime of an inserter depends on the time needed to produce a complete batch and on the lifetime. The downtime depends on the repair and the switch-over times. The parameters are given in Table 2.1.2.

	machine speeds (norm PCBs/ min.)	average uptime (min.)	average downtime (min.)	batch size norm PCBs
horizontal ins.	2.875	32.53	3.61	105
vertical ins.	2.25	43.69	7.11	

Table 2.1.2. Inserter parameters expressed in numbers of norm PCBs

The average uptime of a horizontal inserter in Table 2.1.2 is less than 105/2.875 (the average time to produce a batch) because of machine failures. Given the uptime, the average downtime of an inserter, including change–overs, can be determined, taking the downtime fraction $\frac{\text{downtime}}{\text{uptime+downtime}}$ equal to .10, for a horizontal inserter and equal to .14 for a vertical inserter. See Table 2.1.1.

In the continuous–flow model it is assumed that the inserters do not contain products. In reality an inserter may contain a PCB, but the number of PCBs in the inserters is approximately constant (about 7) and small anyway, compared with the WIP norm. If the number of products in the inserters would be exactly constant, equal to 7, then these 7 products in the inserters would not have a real buffering function. This leads to a problem in the model, since there all K products can be used for buffering against disturbances. The problem can be solved in the model, by subtracting the number of PCBs within the inserters from the WIP norm.

A similar problem (but with even less influence on the performance) arises because in the model vertical insertion keeps on producing if less than a single PCB is available, and horizontal insertion keeps on producing if the buffer content is larger than K–1, due to the continuous product flow. Clearly, this is not true in the real system. This representation error can be mended by subtracting two PCBs from the upper workload norm in the model. Let K' be the WIP norm in the model. The influence of these 9 PCBs in total, not incorporated in the model, is taken into account, by defining K' := K–9 PCBs in the model, where K is the real workload norm.

The model of the insertion department is equivalent, as far as the throughput is concerned, to a two–stage line consisting of horizontal and vertical insertion and an intermediate buffer of physical capacity K'. A content K'–X(t) of the buffer preceding horizontal insertion, and hence X(t) of the intermediate buffer ($0 \leq X(t) \leq K'$) for the insertion department, corresponds to a content X(t) of the intermediate buffer in this two-stage line. See Fig. 2.1.3.

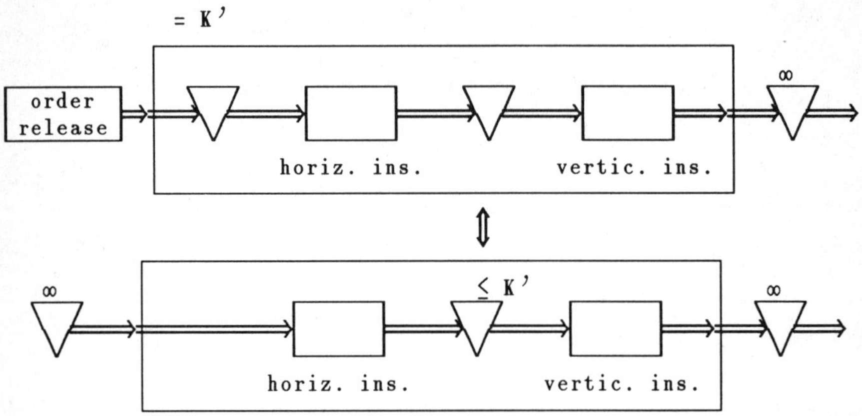

Figure 2.1.3. The insertion department model with constant WIP level K' is equivalent to horizontal and vertical insertion with an intermediate buffer of capacity K'

Note that, due to the continuous modeling, the inserters cannot contain products. That the two lines in Fig. 2.1.3 are equivalent can be understood by observing that horizontal and vertical insertion in the upper line, with buffer contents (K'–X(t),X(t)), respectively, behave in exactly the same way as horizontal and vertical insertion in the lower line with buffer content X(t), for all X(t), with $0 \leq X(t) \leq K'$. In both cases horizontal insertion is blocked if the content of the intermediate buffer equals K', and vertical insertion is starved if the intermediate buffer is empty.

In conclusion, the resulting model can be summarized as follows:

- The two-stage model consists of a line with continuous product flow and a finite intermediate buffer (capacity K')
- Each stage consists of several unreliable machines, each machine with an up and a down state and a constant machine speed.
- The upstream stage always has sufficient components and bare PCBs to work on. The downstream stage always has storage space for finished PCBs.

The model of the department as described here will be analyzed in section 2.9.

Before starting the analysis of this type of model, a survey of existing (analytical) literature on two–stage lines is given in the next section.

2.2 LITERATURE ON TWO–STAGE LINES

In this section we give some related literature on analytic models of two–stage lines with finite buffers. Most of this literature deals with the problem of calculating performance measures like throughput, mean system content and buffer content distribution. Only a rough survey is given here.

There is much literature on analytic models of two–stage lines with finite capacity intermediate buffers. Surveys are given by Koenigsberg (1959), Buzacott and Hanifin (1978), Forestier (1980), Gershwin and Berman (1981), Neuts (1981), Yeralan and Muth (1987), Perros (1986), Gün (1987) and Awate and Sastry (1987). In some of the literature mentioned hereafter longer lines (three or more stages) are also studied. Literature on systems consisting of more than one finite buffer is surveyed in Chapter 3. For such systems hardly any analytic or numeric results exist.

At least four classes of models can be distinguished. The first deals with systems in which the service times are random variables and the products are discrete. The machines are not susceptible to failures. There are three main blocking types, which are defined for general queueing networks (see Onvural and Perros, 1986). All types have various subtypes. The three blocking mechanisms are

- *transfer blocking.* After the product has been completed by machine i, it declares a destination waiting room j. If waiting room j is full at that moment, machine i becomes blocked and cannot operate on other products, for the time that waiting room j is full.
- *communication blocking.* The product to be operated by machine i declares its destination waiting room j before receiving service. If waiting room j is full at that instant, then machine i is blocked and the product does not receive service for the duration of the blocking
- *Repeated transfer blocking.* Similar to transfer blocking, but the product receives another service if the destination queue is full on service completion. This is repeated until the destination queue is not full at the instant of service completion.

These blocking mechanisms and certain of their subtypes have been compared by various authors like Caseau and Pujolle (1979), Altiok and Stidham (1982), Bocharov and Albores (1980) and Onvural and Perros (1986). It has been shown that for many networks, various blocking mechanisms are equivalent, that is, the marginal queue–length probabilities are identical.
Hunt (1956), Hildebrand (1967, 1968), Pujolle and Potier (1979), Neuts (1968, 1970), Konheim and Reiser (1976, 1978) assume there is an unbounded input queue upstream of the first machine and the parts arrive randomly. Other authors assume there is an inexhaustible supply of products to the first machine. See Hillier and Boling (1966, 1967), Hatcher (1969), Knott (1970), Kraemer and Love (1970), Muth (1973, 1977), Foster and Perros (1980) and Rao (1975a, 1975b, 1976)). Service times are often assumed to be exponentially distributed (Hunt (1956), Patterson (1964), Latouche and Neuts (1980), Lavenberg (1978), Wong et al. (1977), Gordon and Newell (1967), Hildebrand (1968), Kraemer and Love (1970), Konheim and Reiser (1976, 1978), Pujolle and Potier (1979)). However, also other distributions have been studied, for instance Erlang distributions by Hillier and Boling (1967) and more general distributions by Rao (1975a, 1975b, 1976), Muth (1973, 1977) and Gün and Makowski (1988).

A second class of models assumes deterministic processing times, but machines are unreliable and fail from time to time. Products are discrete. Examples are Buzacott (1967, 1982), Sheskin (1976), Soyster and Toof (1976), Artamonov (1977), Okamura and Yamashina (1977), Ignall and Silver (1977), Buzacott and Hanifin (1978), Sutalaksana and Van Wassenhove (1981), Sutalaksana et al. (1982), Shantikumar and Tien (1983) and Yeralan and Muth (1987). In these papers (except in Sutalaksana et al. (1981, 1982) and Yeralan and Muth (1987)) life and repair times are geometrically distributed.

Another class of papers deals with continuous flow models. In these papers machine speeds are deterministic, but machines may fail. This class of papers contains old work like Zimmern (1956), Finch, Vladzievskii (see Koenigsberg (1959)), Sevast'yanov (1962) and more recent work of Buzacott and Hanifin (1978), Murphy (1975, 1978, 1979), Fox and Zerbe (1974), Wijngaard (1979), Meyer et al. (1979), Malathronas et al. (1983), Dubois and Forestier (1982), Anick et al. (1982), Coillard and Proth (1984), Hontelez (1985), Mitra (1986, 1988), Glassey and Hong (1986a) and Yeralan et al. (1986).

A fourth class contains papers with discrete products, failures of machines and random processing times. This class contains work of Buzacott (1972), Gershwin and Berman (1981), Klaren (1987) and Buzacott and Kostelski (1987).

The models of classes 2 and 3 are closely related to the model used in this text. Therefore some models (mainly) belonging to these classes, will be compared in more detail. The models differ in assumptions. Some assumptions have a substantial influence on the line performance, other assumptions are less important. All discrete–product models of class 2 assume equal machine–cycle times. A comparison of a number of analytical two–stage models with two unreliable machines and a buffer (of capacity $0<K<\infty$) is given in Table 2.2.1. Comparative characteristics are:

1. the product flow (continuous or discrete).
2. Failure behavior of machines: lifetime distribution (e: exponential, g: geometric); do stations fail at the same rate, when they are slowed down, blocked or starved? (if yes, then the failures are time–dependent, if not then the failures are state–dependent. A distinction can be made between different types of state–dependent failures); can the machines, both being up, fail at the same instant?; can both machines be down at the same instant?
3. Downtime behavior of the machines: downtime distribution; what happens on breakdown (in the discrete product case) with the product the machine was operating? Is it released to the next following buffer in a finished form, is it held by either the preceding buffer or by the machine, or is it (possibly) scrapped?; can machines which have failed be repaired at the same instant?
4. For the discrete–product case (within a machine cycle): is the observation instant prior to the transfer of the product, after failures or repairs, or after product transfer, before failures or repairs?
5. Are the machines in the model identical and if not
6. do the machines have equal cycle times?

Some of the discrete product models have been unified by Yeralan and Muth (1987). In some discrete product models extra machine states are introduced for the case that a machine is blocked or starved (e.g. Okamura and Yamashina (1977) and Buzacott (1972)).

In section 2.3.7 a discrete product model of class 2 is compared with the continuous flow model. It will be shown that they are very close in performance (throughput, mean work–in–process) for a proper choice of parameters.

	1	2	3	4	5	6	7	8	9	10	11	12	13	14	15	16
Flow																
continuous	x	x	x						x				x		x	x
discrete				x	x	x	x	x		x	x	x		x		
Failures																
Uptime distribution	e^2	e	e^2	g^2	g	g	g^2	g^2	e	e	g^2	g^2	e^2	g^2	e^2	e
Type of failures																
state dependent	x		x	x			x	x	x	x	x	x	x	x	x	x
time dependent		x			x^3	x			x						x	
Simult. station failure																
yes					x	x	x	x			x	x		x		
no	x	x	x	x					x	x			x		x	x
Both stations failed																
yes		x		x	x	x	x	x	x	x	x	x	x	x	x	x
no	x		x													
Repairs																
Downtime distribution	e	e	e	e	g	g	g	g	e	e	g	g	e	g	e	e
Upon breakdown[4]																
release					x	x	x			x	x	x				
hold				x												
(possible) scrap								x						x		
Simult. station repair																
yes					x	x	x	x			x	x		x		
no	x	x	x	x					x	x			x		x	x
Observation instant[4]																
before transfer				x	x^5	x^5					x			x		
after transfer					x^5	x^5	x	x		x		x				
Machines																
identical machines																
yes	x			x				x	x			x				
no		x	x		x	x	x	x	x	x	x	x	x	x	x	x
equal cycle times																
yes	x	x	x	x	x	x	x	x	x		x	x	x	x	x	
no									x	x						x

Table 2.2.1. Comparison of two–stage models. 1: Vladzievskii, 2: Finch (see Koenigsberg, 1959), 3: Sevast'yanov (1962), 4: Buzacott (1972), 5: Sheskin (1976), 6: Soyster and Toof (1976), 7: Artamonov (1976), 8: Okamura and Yamashina (1977), 9: Wijngaard (1979), 10 : Gershwin and Berman (1981), 11: Buzacott (1982), 12: Gershwin and Schick (1979), 13: Malathronas et al. (1983), 14: Shantikumar and Tien (1983), 15: Yeralan et al. (1986). 16: Mitra (1986). ^{2}= constant failure rate during the time that both stations are up, ^{2}= measured in working time, ^{3}= failure rate+repair rate = 1, ^{4}= for the discrete product case, ^{5}= two–step transfer.

2.3 THE TWO-STAGE MODEL

2.3.1 Introduction

In section 2.1, modeling the product flow continuously arose in a natural way. In this and the following sections it will be demonstrated how continuous two-stage models can be analyzed and how important variables (like throughput, mean buffer contents, buffer-content distribution) can be calculated. We start the analysis with a simple model of two unreliable machines, with exponentially distributed up and down times (subsections 2.3.2, 2.3.3 and 2.3.4). In following subsections certain assumptions will be dropped subsequently and the adapted model is analyzed.

2.3.2. Model description

The simplest continuous product-flow model is sketched in Fig. 2.3.1.

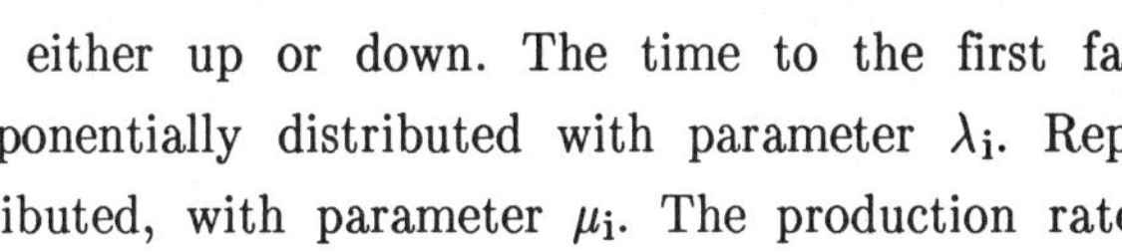

Figure 2.3.1 Two-stage line

The machines are either up or down. The time to the first failure of machine i (M_i, i=1,2) is exponentially distributed with parameter λ_i. Repair times are also exponentially distributed, with parameter μ_i. The production rates of the machines are constant. The buffer capacity is denoted by K and machine speeds by ν_i, i=1,2.

Due to the finiteness of the buffer capacity, the machines may be blocked or starved. Blocking and starvation are special cases of slowing down. If the buffer is full and the instantaneous machine speed of M_2 is less than that of M_1, then M_1 has to slow down to the speed of M_2. Hence, if the buffer is full, M_1 is up and M_2 is down, then M_1 slows down to rate 0. M_1 is then said to be *blocked.* If the buffer is empty and the instantaneous rate of M_2 is greater than that of M_1, then M_2 has to slow down until both rates are equal. M_2 may become completely *starved* if B is empty and M_1 is down. It is assumed that M_1 is never slowed down by lack of input and M_2 is never slowed down by lack of storage capacity for output. In what follows, similar assumptions will be made for more complicated networks of machines and buffers. It is assumed that the failure rates of the machines are *time-dependent*, rather than state-dependent, i.e. failures occur at the same rate when machines are slowed down. This particular model is analyzed in subsections 2.3.3 and 2.3.4. In the following subsections we will subsequently drop certain assumptions made previously and analyze the adapted model. For completeness we give a list of assumptions made here, but discussed later. The subsection where the adapted model is discussed is indicated between parentheses.

(a) Failures are time-dependent (state-dependent failures in subsection 2.3.6).
(b) Machines have two states, up and down, each with exponential sojourn time (phase-type distributions and groups of parallel machines in subsection 2.4).
(c) The buffer has finite capacity and backlogging is not allowed (infinite buffers and backlogging in subsection 2.5.1).
(d) Machine M_1 can only be blocked or slowed down if the buffer is full (different buffer-operating policies in subsection 2.5.2).

2.3.3 Model analysis

In this subsection the two-stage line is analyzed following the method of Wijngaard (1979).

The machines of the two-stage line can be in four different machine states, namely:
1: M_1 up, M_2 down,
2: M_1 down, M_2 up,
3: M_1 down, M_2 down,
4: M_1 up, M_2 up.
The state of the system is determined by (i,x), where $i \in \{1,2,3,4\}$ denotes the machine state and $x \in [0,K]$ denotes the buffer content. We distinguish so-called *regeneration points*, points in time where the process probabilistically restarts itself (see e.g. Ross, 1970), for instance an entrance to (2,0). That is, the machines are in state 2 and the buffer is empty. Note that this regeneration point can be chosen only if $\lambda_1 \neq 0$, in other words M_1 is not perfect. If $\lambda_1=0$ and $\lambda_2 \neq 0$ an entrance to (1,K) can be chosen. Unless stated otherwise we suppose for the rest of this chapter that $\lambda_1 \neq 0$ and the regeneration point is (2,0).

The most important performance quantities are the throughput, the average buffer content and the average throughput time.

The *throughput* v(K) is defined as the average (long run) line production rate, or $\lim_{t\to\infty} \int_0^t w(\tau)d\tau/t$, where $w(\tau)$ stands for the line output rate at time τ.
The *average (long-run) buffer content* EX is defined as $\lim_{t\to\infty} \int_0^t X(\tau)d\tau/t$, where $X(\tau)$ stands for the buffer content at time τ.

The *average (long-run) throughput time* S(K) can be obtained from the throughput and average buffer content by Little's formula, EX = v(K)S(K). Other quantities of interest are the buffer-content probability distribution function and the blocking and starvation probabilities. To obtain these quantities we can argue as follows. For the throughput, for example, the mean production per cycle P and the mean cycle length T are calculated. A cycle is the time between two subsequent regenerations. By a renewal theoretical argument, it follows that the throughput is then equal to P/T (see Ross, 1970, proposition 5.9). Other performance quantities can be obtained in the same way. Quantities as P and T can be interpreted as "costs" per cycle, where the word "cost" is used as having a common denominator for all possible contributions to the particular quantity in a cycle, rather than for optimization purposes. All these cycle costs can be determined in a similar way.

Let $\alpha_i(x)$ be the cost rate in state i with inventory level $x \in [0,K]$. To determine cycle costs we define $f_i(x)$ as the expected cost till the end of the cycle (that is till the first entrance to (2,0)) if we are now in machine state i, with inventory level x. From the physical interpretation it follows that the costs f_i are continuous. For the choice of (2,0) as regeneration point, the cost f_2 is not defined in 0. To make f_2 also continuous in 0, $f_2(0)$ is defined to be 0.

Summarizing, the assumptions are:
1. The regeneration point is (2,0)
2. $f_2(0) = 0$.

The expected cost per cycle, C_T, can now be written as

$$C_T = \frac{1}{\lambda_2+\mu_1}[\alpha_2(0) + \lambda_2 f_3(0) + \mu_1 f_4(0)] \tag{2.1}$$

By splitting the cost till the end of the cycle in the cost during the first small time period Δ and the cost during the rest of the cycle, the following equation for $f_1(x)$ can be obtained, for $0 \le x < K$.

$$f_1(x) = \alpha_1(x)\Delta + (1-\lambda_1\Delta)(1-\mu_2\Delta)f_1(x+\nu_1\Delta) + \lambda_1\Delta f_3(x+o(\Delta)) + \mu_2\Delta f_4(x+o(\Delta)) + o(\Delta).$$

by dividing by Δ and letting $\Delta \to 0$, we obtain

$$-\nu_1 f_1'(x) = \alpha_1(x)-(\lambda_1+\mu_2)f_1(x)+\lambda_1 f_3(x)+\mu_2 f_4(x) , \qquad 0 \le x < K \qquad (2.2)$$

Differential equations for the other f_i can be obtained similarly (see also Wijngaard, 1979)

$$\nu_2 f_2'(x) = \alpha_2(x)-(\lambda_2+\mu_1)f_2(x)+\lambda_2 f_3(x)+\mu_1 f_4(x) , \qquad 0 < x \le K \qquad (2.3)$$

$$0 = \alpha_3(x)-(\mu_1+\mu_2)f_3(x)+\mu_1 f_1(x)+\mu_2 f_2(x) , \qquad 0 \le x \le K \qquad (2.4)$$

$$(\nu_2-\nu_1)f_4'(x) = \alpha_4(x)-(\lambda_1+\lambda_2)f_4(x)+\lambda_1 f_2(x)+\lambda_2 f_1(x) , \qquad 0 < x < K \qquad (2.5)$$

If $\nu_1 > \nu_2$, then equation (2.5) also holds for $x=0$. If $\nu_1 < \nu_2$, then equation (2.5) also holds for $x=K$. The boundary conditions are obtained from these differential equations, by observing that at the boundary point the buffer filling or emptying rate of a machine has to be adapted. For instance, for $x=K$ in state 1, M_1 is blocked and hence $\nu_1=0$. This yields boundary equation (2.2a). The other boundary equations can be obtained similarly.
For $\nu_1 > \nu_2$ we obtain,

$$0 = \alpha_1(K) - (\lambda_1+\mu_2)f_1(K) + \lambda_1 f_3(K) + \mu_2 f_4(K) \qquad (2.2a)$$

$$0 = f_2(0) \qquad (2.3a)$$

$$0 = \alpha_4(K) - (\lambda_1+\lambda_2)f_4(K) + \lambda_1 f_2(K) + \lambda_2 f_1(K) \qquad (2.5a)$$

For $\nu_2 \ge \nu_1$ the latter boundary condition has to be replaced by

$$0 = \alpha_4(0) - (\lambda_1+\lambda_2)f_4(0) + \lambda_1 f_2(0) + \lambda_2 f_1(0) \qquad (2.5b)$$

In matrix notation the system of differential equations (2.2)–(2.5) can be rewritten as

$$\frac{\partial}{\partial x}(-\mathbf{D})\mathbf{f}(x) = \mathbf{M}\mathbf{f}(x) + \boldsymbol{\alpha}(x) \tag{2.6}$$

The matrix **D** is called the *drift matrix*, it is diagonal with diagonal entries $(\nu_1,-\nu_2,0,\nu_1-\nu_2)$. The matrix **M** is the generator of the Markov process of the machine states.

The system (2.6) can be solved. Equation (2.4) can be substituted in the other ones, so that a system of three differential equations and three boundary equations remains. In section 2.3.4 it will appear that for the general case of $\nu_1 \neq \nu_2$ and $\lambda_1 \neq 0$ always a homogeneous solution can be found. It can also be shown that exactly one solution of the inhomogeneous system, satisfying the boundary conditions, exists.

By substituting the right values for the α_i's in (2.1)–(2.5b) all line characteristics of interest can be calculated from equation (2.1). In Table 2.3.1 a list is given of some interesting quantities which can be calculated. Not all these quantities are needed in this chapter. In this chapter only the throughput, average buffer content and the buffer content distribution are calculated. For these quantities P, T, Y and PL are needed. The other quantities in Table 2.3.1 are used in chapters 4 and 5, for the combination of different machines. They are defined here, since they all can be obtained in a similar way.

Symbol	description
T	average cycle length
P	average production, per cycle
I_ν	average time, per cycle, the input rate of the line is equal to ν
S_ν	average time, per cycle, the output rate of the line is equal to ν
O_i	average number of times per cycle that the line has input, or average number of times per cycle that the line has no input
O_o	average number of times per cycle that the line has output, or average number of times per cycle that the line has no output
Q_1	average time per cycle that M_1 is blocked
Q_2	average time per cycle that M_2 is starved
Y	average buffer content
PL	average time, per cycle, that the buffer content is greater than, or equal to L ($L \in [0,K]$)

Table 2.3.1. Some line quantities for the two-stage line

	$\alpha_1(x)$	$\alpha_2(x)$	$\alpha_3(x)$	$\alpha_4(x)$	$\alpha_1(K)$	$\alpha_2(0)$	$\alpha_3(0)$	$\alpha_3(K)$	$\alpha_4(0)$	$\alpha_4(K)$
T	1	1	1	1	1	1	1	1	1	1
P	0	ν_2	0	ν_2	0	0	0	0	$\min\{\nu_1,\nu_2\}$	ν_2
I_{ν_1}	1	0	0	1	0	0	0	0	1	$\delta_{\{\nu_1\le\nu_2\}}$
I_{ν_2}	0	0	0	0	0	0	0	0	0	$\delta_{\{\nu_2<\nu_1\}}$
I_0	0	1	1	0	1	1	1	1	0	0
S_{ν_1}	0	0	0	0	0	0	0	0	$\delta_{\{\nu_1<\nu_2\}}$	0
S_{ν_2}	0	1	0	1	0	0	0	0	$\delta_{\{\nu_2\le\nu_1\}}$	1
S_0	1	0	1	0	1	1	1	1	0	0
O_i	0	μ_1	μ_1	0	μ_2	μ_1	μ_1	0	0	0
O_o	μ_2	0	μ_2	0	μ_2	μ_1	0	μ_2	0	0
Q_1	0	0	0	0	1	0	0	0	0	0
Q_2	0	0	0	0	0	1	0	0	0	0
Y	x	x	x	x	K	0	0	K	0	K
PL	$\delta_{\{x\ge L\}}$	$\delta_{\{x\ge L\}}$	$\delta_{\{x\ge L\}}$	$\delta_{\{x\ge L\}}$	1	$\delta_{\{0\ge L\}}$	$\delta_{\{0\ge L\}}$	1	$\delta_{\{0\ge L\}}$	1

Table 2.3.2. Determination of the line quantities in Table 2.3.1.
$\delta_{\{x\}}$, for some condition x, stands for $\delta_{\{x\}} = \begin{cases} 1, & \text{if x holds} \\ 0, & \text{if x does not hold} \end{cases}$

The determination of O_i and O_o may require an explanation. For these cases, the values for α_i represent transition rates from a state without input or output, respectively, to a state with input or output, respectively. For instance, μ_1 in state (2,x) is the rate with which a jump is made from a state without input, to state (4,x), i.e. a state with input.

From the numbers in Table 2.3.2 it will be clear that there are relations between the quantities listed in Table 2.3.1. For instance, PL, with L equal to 0, equals the cycle time T. For the throughput we have

$$v(K) = \frac{P}{T} = \frac{\nu_1 S_{\nu_1} + \nu_2 S_{\nu_2}}{S_0 + S_{\nu_1} + S_{\nu_2}} = \frac{\nu_1 I_{\nu_1} + \nu_2 I_{\nu_2}}{I_0 + I_{\nu_1} + I_{\nu_2}}$$

The mean buffer content EX, equals Y/T and $P[X \le L] = (T-PL)/T$ ($L \in [0,K]$).

The eventual objective is to determine the quantities in Table 2.3.1. Such a quantity is determined as follows. For each variable in Table 2.3.1, the α_i are determined as indicated in Table 2.3.2. These values for the α_i are substituted in the system (2.1)–(2.5b). Then the system of differential equations is solved. This yields the values of the functions f_i in 0. The desired quantity is now obtained from C_T, by substituting these values in the cost formula (2.1). Hence, the problem is to solve the system (2.2)–(2.5b).

The system of differential equations and boundary equations (2.2)–(2.5b) becomes simpler if any of the machines is perfect, or if the machine speeds are equal. For these cases also the solution technique differs. Hence the following cases have to be distinguished: (a) $\lambda_1=0$ or $\lambda_2=0$ (in case $\lambda_2=0$ also the regeneration point changes), (b) $\nu_1=\nu_2$, (c) $\lambda_1\neq 0$, $\lambda_2\neq 0$, $\nu_1\neq\nu_2$. Cases (a) and (b) are special, they can be solved analytically. Case (c) is the most general. It is solved numerically. In the next subsection case (c) is treated. Cases (a) and (b) are treated in Appendix A.

2.3.4 Solution of the model for $\lambda_1\neq 0$, $\lambda_2\neq 0$ and $\nu_1\neq\nu_2$

In this section we suppose that both machines are unreliable and the machine speeds are unequal.

Substitution of equation (2.4) in (2.2)–(2.5) yields

$$\mathbf{h}'(x) = \mathbf{c}(x) + \mathbf{A}\mathbf{h}(x), \quad 0<x<K \quad , \tag{2.7}$$

where $\mathbf{h}(x)$ is the vector $(f_1(x),f_2(x),f_4(x))^T$, $\mathbf{c}(x)$ is the vector

$$\left(\frac{1}{\nu_1}[-\alpha_1(x) - \frac{\lambda_1}{\mu_1+\mu_2}\alpha_3(x)],\ \frac{1}{\nu_2}[\alpha_2(x) + \frac{\lambda_2}{\mu_1+\mu_2}\alpha_3(x)],\ \frac{1}{\nu_2-\nu_1}\alpha_4(x)\right)^T \tag{2.7a}$$

and $\mathbf{A}$ is the matrix

$$\begin{bmatrix} \frac{1}{\nu_1}\left[\mu_2 + \frac{\lambda_1\mu_2}{\mu_1+\mu_2}\right] & \frac{-\lambda_1\mu_2}{\nu_1(\mu_1+\mu_2)} & \frac{-\mu_2}{\nu_1} \\ \frac{\lambda_2\mu_1}{\nu_2(\mu_1+\mu_2)} & \frac{-1}{\nu_2}\left[\mu_1 + \frac{\lambda_2\mu_1}{\mu_1+\mu_2}\right] & \frac{\mu_1}{\nu_2} \\ \frac{\lambda_2}{\nu_2-\nu_1} & \frac{\lambda_1}{\nu_2-\nu_1} & \frac{-(\lambda_1+\lambda_2)}{\nu_2-\nu_1} \end{bmatrix} \tag{2.7b}$$

The boundary conditions can be rewritten as

$$\alpha_1(K) + \frac{\lambda_1}{\mu_1+\mu_2}\alpha_3(K) - (\mu_2 + \frac{\lambda_1\mu_2}{\mu_1+\mu_2})\,h_1(K) + \frac{\lambda_1\mu_2}{\mu_1+\mu_2}\,h_2(K) + \mu_2 h_3(K) = 0$$

$$h_2(0) = 0$$

$$\alpha_4(K) + \lambda_2 h_1(K) + \lambda_1 h_2(K) - (\lambda_1+\lambda_2)h_3(K) = 0, \qquad \text{if } \nu_2>\nu_1$$

$$\alpha_4(0) + \lambda_2 h_1(0) + \qquad - (\lambda_1+\lambda_2)h_3(0) = 0, \qquad \text{if } \nu_2<\nu_1 \tag{2.7c}$$

The cost per cycle is

$$C_T = \frac{1}{\lambda_2+\mu_1}[\alpha_2(0) + \frac{\lambda_2}{\mu_1+\mu_2}\alpha_3(0) + \frac{\lambda_2\mu_1}{\mu_1+\mu_2}h_1(0) + \mu_1 h_3(0)] \tag{2.8}$$

The problem is to solve the system of differential equations (2.7)–(2.7b), with boundary equations (2.7c). Quantities that will appear to be of interest in the solution method are

The *efficiency* of a machine, defined as $\text{eff}_i := \frac{\nu_i\mu_i}{\lambda_i+\mu_i}$. This is the average stand–alone machine speed of M_i.
The *mean drift* σ of the system. $\sigma := \text{eff}_1 - \text{eff}_2$.

The solution $\mathbf{h}(x)$ of (2.7) is the sum of the homogeneous solution $\mathbf{u}(x)$ and a particular solution $\mathbf{v}(x)$.

The form of the *homogeneous solution* depends on the eigenvalues of **A**. Mitra (1986) has extensively analyzed the eigenvalues of drift processes as described by equation (2.6). By applying his results to our specific problem it follows immediately that all eigenvalues of **A** are real. At least one of the eigenvalues of **A** is 0 and the corresponding right eigenvector is $(1,1,1)^T$. Mitra has shown that the sign of the eigenvalues depends on σ, the *mean drift* of the system. The line is called *balanced* if $\sigma=0$ and *unbalanced* if $\sigma\neq0$. The resulting subdivision in sign of of the eigenvalues of **A** is given in Table 2.3.3.

	# positive eigenvalues	multiplicity of eigenvalues at 0	# negative eigenvalues
$\sigma < 0$	$2\delta_{\{\nu_1>\nu_2\}} + \delta_{\{\nu_1<\nu_2\}}$	1	$\delta_{\{\nu_1<\nu_2\}}$
$\sigma = 0$	$\delta_{\{\nu_1>\nu_2\}}$	2	$\delta_{\{\nu_1<\nu_2\}}$
$\sigma > 0$	$\delta_{\{\nu_1>\nu_2\}}$	1	$\delta_{\{\nu_1>\nu_2\}} + 2\delta_{\{\nu_1<\nu_2\}}$

Table 2.3.3. Sign of the eigenvalues, depending on the mean drift

The general solution of the homogeneous equation is, in case $\sigma \neq 0$

$$\mathbf{u}(x) = C_1\mathbf{e}_1 e^{\rho_1 x} + C_2\mathbf{e}_2 e^{\rho_2 x} + C_3\mathbf{e}_3 e^{\rho_3 x}, \tag{2.9}$$

with C_1, C_2 and C_3 arbitrary constants and ρ_1, ρ_2 and ρ_3 the different eigenvalues of **A** with corresponding eigenvectors $\mathbf{e}_1$, $\mathbf{e}_2$ and $\mathbf{e}_3$.

For the balanced line, that is, the line with $\sigma=0$, the general solution of the homogeneous equation is

$$\mathbf{u}(x) = C_1\mathbf{e}_1 + C_2(\mathbf{e}_2^* + x\mathbf{e}_1) + C_3\mathbf{e}_3 e^{\rho_3 x}, \tag{2.10}$$

where C_1, C_2 and C_3 are again arbitrary constants, $\mathbf{e}_1 = (1,1,1)^T$, ρ_3 is the eigenvalue unequal to 0, $\mathbf{e}_3$ is the corresponding eigenvector and $\mathbf{e}_2^*$ is a *generalized eigenvector of order 2* (see Pease, 1965, p.77). That is, it satisfies

$$\mathbf{A}\mathbf{e}_2^* = \mathbf{e}_1.$$

The *inhomogeneous system.* For the determination of the quantities T (average cycle time), P (average production, per cycle), Y (average buffer content) and PL (average time per cycle that the buffer content is greater than or equal to L), the $\alpha_i(x)$ are determined by the corresponding values in Table 2.3.2. Depending on the values of the $\alpha_i(x)$, the inhomogeneous part $\mathbf{c}(x)$ in equation (2.7) changes. The cost vector $\mathbf{c}(x)$ depends on the $\alpha_i(x)$ through (2.7a). $\mathbf{c}(x)$ is either constant (for T, P, PL) or linear in x (for Y). For T and P we have $\mathbf{c}(x) = \mathbf{c}$. For Y we have $\mathbf{c}(x) = \mathbf{c}x$, for PL we have $\mathbf{c}(x) = \mathbf{c}$ for $x \geq L$ and $\mathbf{c}(x) = \mathbf{0}$ for $x<L$.

First we consider *the case $\mathbf{c}(x)=\mathbf{c}$,* for all $x \in (0,K)$ (determination of T, P).
In case $\alpha_i(x)=\alpha_i$ (independent of x) and hence $\mathbf{c}(x)=\mathbf{c}$ in (2.7), a *standard inhomogeneous system* is obtained. The solution of such systems is treated in Coddington and Levinson (1955). A particular solution of (2.7) is

$$\mathbf{v}(x) = \beta(1,1,1)^T x + \mathbf{d} \tag{2.11}$$

the first component of $\mathbf{d}$, d_1, can be chosen equal to 1. β is a scalar, such that

$$\beta(1,1,1)^T = \mathbf{c} + \mathbf{A}\mathbf{d}$$

This last system of equations yields β, d_2 and d_3.

Combination of the equations (2.9) or (2.10) and (2.11) yields the general solution $\mathbf{h}(x) = \mathbf{u}(x)+\mathbf{v}(x)$ of the system (2.7). The constants C_i in $\mathbf{u}(x)$ are still unknown. They are determined by substituting $\mathbf{h}(x) = \mathbf{u}(x) + \mathbf{v}(x)$ in the boundary equations (2.7c). C_T is now determined by substitution of $h_1(0) = u_1(0)+v_1(0)$ and $h_3(0) = u_3(0)+v_3(0)$ in (2.8).

Second we consider *the case* $\boldsymbol{c}(x)=\boldsymbol{c}$, $x \geq L$, $\boldsymbol{c}(x)=\boldsymbol{0}$, $x<L$ (determination of PL). The system (2.7) reduces to

$$\begin{aligned} \mathbf{h}'(x) &= \mathbf{c} + A\mathbf{h}(x) \ , \quad x \geq L \\ \mathbf{h}'(x) &= \phantom{\mathbf{c} + {}} A\mathbf{h}(x) \ , \quad x<L \end{aligned} \tag{2.12}$$

The solution of the equations (2.12) for $x \geq L$ is $\mathbf{u}(x) + \mathbf{v}(x)$, where $\mathbf{u}(x)$ is given by (2.9) for $\sigma \neq 0$, and by (2.10) if $\sigma=0$, and $\mathbf{v}(x)$ is given by (2.11). The solution for $x<L$ is $\mathbf{u}^*(x)$, where

$$\mathbf{u}^*(x) = C_1^*\mathbf{e}_1 + C_2^*\mathbf{e}_2 e^{\rho_2 x} + C_3^*\mathbf{e}_3 e^{\rho_3 x} \quad , \quad \sigma \neq 0 \tag{2.13}$$

$$\mathbf{u}^*(x) = C_1^*\mathbf{e}_1 + C_2^*(\mathbf{e}_2^* + x\mathbf{e}_1) + C_3^*\mathbf{e}_3 e^{\rho_3 x} \quad , \quad \sigma=0 \tag{2.14}$$

In (2.13) and (2.14) C_1^*, C_2^* and C_3^* are constants. The numbers ρ_2 and ρ_3 are eigenvalues unequal to 0, with eigenvectors $\mathbf{e}_2$ and $\mathbf{e}_3$, respectively. The eigenvector $\mathbf{e}_1$ is defined as $(1,1,1)^T$. In (2.14) the vector $\mathbf{e}_2^*$ is again a generalized eigenvector of order 2, such that

$$A\mathbf{e}_2^* = \mathbf{e}_1.$$

The constants C_1, C_2, C_3 in the formula for $\mathbf{u}(x)$ and the constants C_1^*, C_2^* and C_3^* are determined by substituting

$$\mathbf{h}(x) = \begin{cases} \mathbf{u}(x) + \mathbf{v}(x) & , \ x \geq L \\ \mathbf{u}^*(x) & , \ x<L \end{cases}$$

in the boundary equations (2.7c) and by requiring $\mathbf{u}^*(L) = \mathbf{u}(L) + \mathbf{v}(L)$. This last condition follows from the continuity of $\mathbf{h}$ in L. Now the function $\mathbf{h}(x)$ is determined completely and PL is obtained by substituting $h_1(0)$ and $h_3(0)$ in (2.8).

Finally we consider *the case* $c(x) = cx$ (determination of Y).
A particular solution of (2.7) is

$$\mathbf{v}(x) = \gamma(1,1,1)^T x^2 + \mathbf{a}x + \mathbf{b}, \tag{2.15}$$

where γ is a scalar. There are seven unknowns in equation (2.15), namely γ, a_1, a_2, a_3, b_1, b_2 and b_3. By choosing $b_1 = 1$ as a normalization and by substituting (2.15) in (2.7) it follows that furthermore

$$2\gamma(1,1,1)^T - \mathbf{c} = \mathbf{A}\mathbf{a}$$
$$\mathbf{a} = \mathbf{A}\mathbf{b}$$

These six equations determine the other six unknowns. The solution of (2.7) is $\mathbf{h}(x) = \mathbf{u}(x) + \mathbf{v}(x)$, where $\mathbf{u}(x)$ is obtained from (2.9) or (2.10) and C_1, C_2, C_3 are determined by substituting $\mathbf{h}(x)$ in the boundary conditions.
Y is now obtained by substituting $h_1(0)$ and $h_3(0)$ in (2.8).

step 1.	Substitute the values of $\alpha_i(x)$ obtained from Table 2.3.2, corresponding with the desired variable, in (2.7a), (2.7c) and (2.8). This yields $\mathbf{c}(x) = \mathbf{c}$.
step 2.	Calculate the eigenvalues $\rho_1:=0,\rho_2,\rho_3$ and the right eigenvectors $\mathbf{e}_1:=(1,1,1)^T,\mathbf{e}_2$ (or $\mathbf{e}_2^*$ satisfying $\mathbf{A}\mathbf{e}_2^* = \mathbf{e}_1$, if $\sigma=0$) and $\mathbf{e}_3$, respectively of the matrix $\mathbf{A}$. See (2.7b).
step 3.	Define $\mathbf{u}(x) := \begin{cases} C_1\mathbf{e}_1+C_2\mathbf{e}_2e^{\rho_2 x}+C_3\mathbf{e}_3e^{\rho_3 x} & ,\ \sigma\neq 0 \\ C_1\mathbf{e}_1+C_2(\mathbf{e}_2^*+x\mathbf{e}_1)+C_3\mathbf{e}_3e^{\rho_3 x} & ,\ \sigma=0 \end{cases},$ with C_1,C_2,C_3 arbitrary constants.
step 4.	Determine the scalar β and the vector $\mathbf{d}$, with $d_1=1$, from the equations $\beta(1,1,1)^T = \mathbf{c} + \mathbf{A}\mathbf{d}$.
step 5.	Define $\mathbf{v}(x) := \beta(1,1,1)^T x + \mathbf{d}$.
step 6.	Define $\mathbf{h}(x) := \mathbf{u}(x)+\mathbf{v}(x)$. Determine the constants C_1,C_2,C_3 by substituting $\mathbf{h}(x)$ in (2.7c). This yields $\mathbf{h}(x)$ and hence $\mathbf{h}(0)$.
step 7.	Substitute $h_1(0)$ and $h_3(0)$ in (2.8). This yields the desired variable.

Table 2.3.4. Scheme for the determination of the quantities in Table 2.3.1 (except PL, Y)

To summarize the results obtained in this section, a scheme is given in Table 2.3.4, by which all variables in Table 2.3.1, except PL and Y, can be determined. It is easy to construct a similar scheme for the determination of PL and Y.

2.3.5. Numeric results

To demonstrate the effect of the buffer size on throughput and mean buffer content of a two–stage line some cases are examined for unbalanced lines ($\sigma \neq 0$). The cases are taken from Wijngaard (1979), with some extensions. The effect of the buffer size on the throughput is given in Figures 2.3.2. We distinguish the cases

a. $\lambda_2 > \lambda_1 = 0.01$, $\mu_2 = \mu_1 = 0.09$, $\nu_2 = \nu_1 = 1$, λ_2 varying

b. $\lambda_1 = \lambda_2 = 0.01$, $\mu_2 < \mu_1 = 0.09$, $\nu_2 = \nu_1 = 1$, μ_2 varying

c. $\lambda_1 = \lambda_2 = 0.01$, $\mu_1 = \mu_2 = 0.09$, $\nu_2 < \nu_1 = 1$, ν_2 varying

d. $\lambda_1 = 0.01$, $\lambda_2 = 0$, $\mu_1 = 0.09$, $\nu_2 < \nu_1 = 1$, ν_2 varying

The net production rate of M_1 is $\frac{\nu_1 \mu_1}{\lambda_1 + \mu_1} = 0.9$. The parameters of M_2 are chosen such that the cases a, b, c and d can be compared for the same net production rate (eff_2). In Table 2.3.5 the parameter choices for the different cases are given. In Figs. 2.3.2, the graphs for $\text{eff}_2 = 0.891$ are given, for $\text{eff}_2 = 0.855$, for $\text{eff}_2 = 0.750$ and for $\text{eff}_2 = 0.600$. In all these pictures it appears that, for a fixed buffer, case d gives the highest throughput, followed by case c, case a and case b. Case d has the smallest variance in machine speed of M_2, followed by case c. Case a is favorable over case b, since many small failures are better than few large ones (given that the efficiency and machine speeds are the same). See also Wijngaard (1979).

	eff₂											
	0.891			0.855			0.750			0.600		
	λ_2	μ_2	ν_2	λ_2	μ_2	ν_2	λ_2	μ_2	ν_2	λ_2	μ_2	ν_2
a.	0.0110	0.09	1	0.0153	0.09	1	0.03	0.09	1	0.06	0.09	1
b.	0.01	0.0817	1	0.01	0.059	1	0.01	0.030	1	0.01	0.015	1
c.	0.01	0.09	0.99	0.01	0.09	0.95	0.01	0.09	0.833	0.01	0.09	0.667
d.	0	–	0.891	0	–	0.855	0	–	0.891	0	–	0.600

Table 2.3.5 Parameters of some two–stage lines

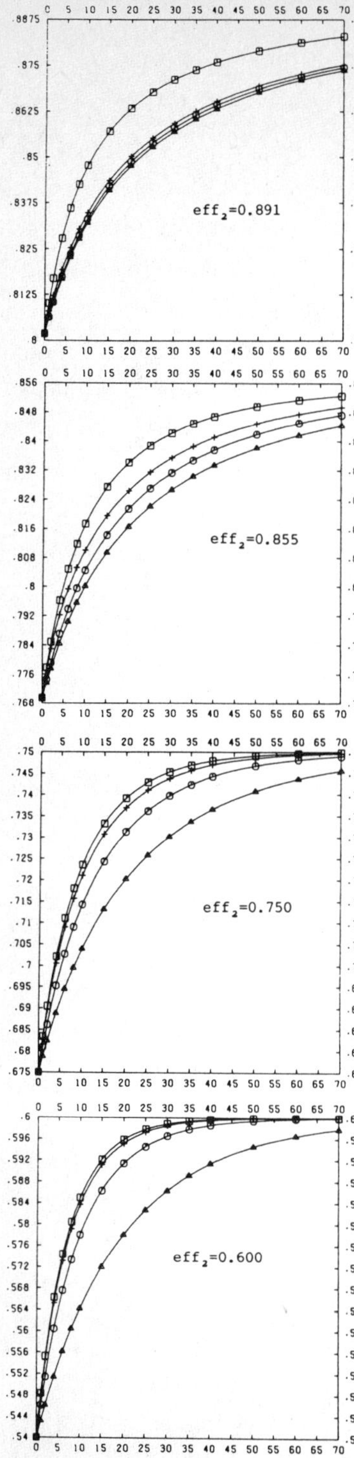

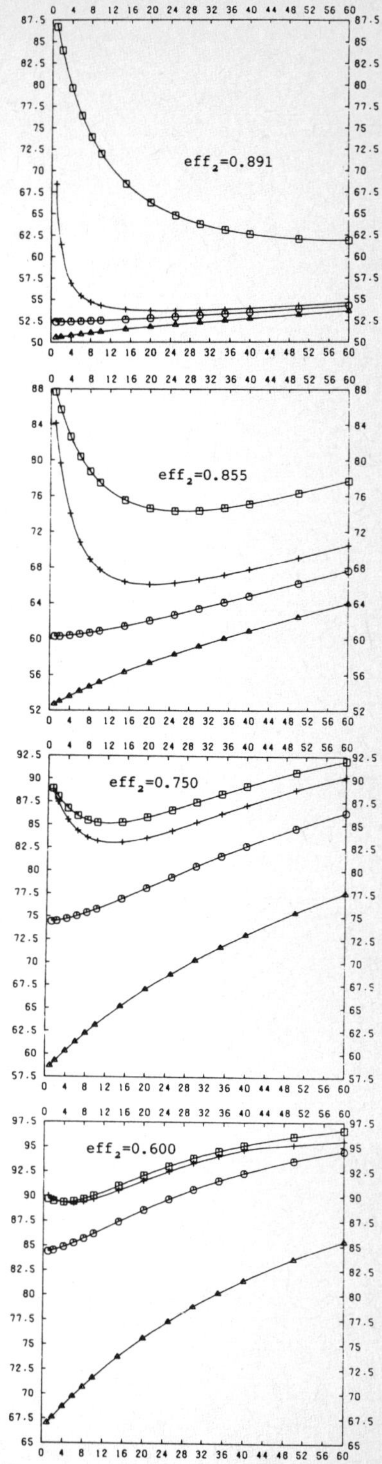

Figures 2.3.2. Line throughput. Figures 2.3.3. Buffer filling percentage. On the horizontal axis the buffer capacity K is shown. The efficiency of M_2 is 0.891, 0.833, 0.750 and 0.600, respectively. □: case d, +:case c, o: case a, Δ: case b.

In Figs. 2.3.3 the filling percentage of the buffer ((EX/K)) × 100%) is given for the same parameters as the throughput.

To conclude this section, an indication is given of the computation time needed for the evaluation of all quantities in Table 2.3.2. In Table 2.3.6 three different methods are compared for CPU time on a Burroughs B7900 computer, for some two-stage lines. Evaluated were all quantities listed in Table 2.3.2. D02GBF stands for the routine D02GBF, for solving general systems of first order differential equations, from the NAG library (1987). DMUTSS stands for the routine DMUTSS of Matthey and Staarink (1985) and SFI stands for straightforward implementation of the method described in section 2.3.4. Each method has its own benefits and drawbacks. D02GBF and DMUTSS are generally applicable. SFI requires a different routine for different situations ($\sigma \neq 0$ and $\sigma = 0$, different inhomogeneous parts), but is faster than the other two.

CPU times	K	D02GBF	DMUTSS	SFI
λ = (0.01,0.01)	2	0.9	1.0	0.5
μ = (0.09,0.09)	5	1.1	1.2	0.5
ν = (1,2)	10	1.1	1.8	0.5
	20	1.6	2.6	0.5
	30	1.5	3.5	0.6
λ = (0.01,0.01)	2	1.0	0.7	0.5
μ = (0.09,0.09)	5	1.1	0.8	0.5
ν = (2,1)	10	1.2	1.3	0.6
	20	1.7	1.9	0.5
	30	2.0	2.5	0.5

Table 2.3.6. CPU times in seconds for three different methods.

2.3.6. State-dependent failures

In the model of section 2.3.2, failures are assumed to be *time-dependent*, that is, they occur at the same rate when a station is blocked or starved, or forced to run at a lower speed. Time-dependent failures will occur in highly automated environments, like chip fabrication, where a constant fall out of dust causes process failures. In production systems also other types of failure behavior can be observed. In some situations machines cannot break down when blocked or starved, or they break down at a faster rate when the machine has to operate faster. This is the case when failure behavior is mainly caused by tool wear. In such situations the failure behavior is called *state–dependent.*

For a discussion on the background of time and state–dependent failures, see Buzacott and Hanifin (1978). State–dependent failures for continuous–flow two–stage lines were treated by Vladzievskii (see Koenigsberg, 1959), Sevast'yanov (1962), Wijngaard (1979), Dubois and Forestier (1982), Malathronas et al. (1983), Yeralan et al. (1986), Hontelez (1985) and Mitra (1986).

We distinguish two notions of state–dependent failures. In both notions, the failure rate is constant as long as machines are not blocked or starved. The failure rate of a machine is 0 when it is blocked or starved. In the second notion of state–dependent failures, it is furthermore assumed that the failure rates are adapted for machines that are slowed down. Note that in case $\nu_1=\nu_2$, the two notions of state–dependent failures are the same.

For both notions of state–dependent failure behavior it is straightforward to adapt the analysis in section 2.3 to the particular notion of state dependent–failure behavior. By assuming that the failure rate is adapted in blocked, starved or slowed–down states, only the boundary conditions (2.2a)–(2.5b) and the cost function (2.1) have to be changed. The system (2.2)–(2.5) remains unchanged. This implies, in case $\nu_1\neq\nu_2$, that the general solution of the homogeneous system is also given by (2.9) or (2.10) and that the particular solution will also be the same as in subsection 2.3.4. Only the constants C_1, C_2 and C_3 are different. These constants depend on the boundary conditions.

We treat as an example the second notion of state–dependent failure behavior, as discussed by Wijngaard (1979). Wijngaard assumes the failure rate to be 0 for a machine that is blocked or starved. He also assumes that in state (4,K) the failure rate of M_1 is $\min\{\nu_1,\nu_2\}\cdot\lambda_1/\nu_1$ and in state (4,0) the failure rate of M_2 is $\min\{\nu_1,\nu_2\}\cdot\lambda_2/\nu_2$. Hence the failure rate is also adapted if the machine is slowed down.

Under Wijngaard's assumptions, instead of (2.1), the cost becomes

$$C_T = \frac{\alpha_2(0)}{\mu_1} + f_4(0) \tag{2.16}$$

The boundary conditions (2.2a), (2.5a) and 2.5b) have to be replaced by

$$0 = \alpha_1(K) - \mu_2 f_1(K) + \mu_2 f_4(K) \tag{2.16a}$$

$$0 = \alpha_4(K) - (\frac{\lambda_1 \nu_2}{\nu_1} + \lambda_2) f_4(K) + \frac{\lambda_1 \nu_2}{\nu_1} f_2(K) + \lambda_2 f_1(K) \tag{2.16b}$$

$$0 = \alpha_4(0) - (\lambda_1 + \frac{\lambda_2 \nu_1}{\nu_2}) f_4(0) + \lambda_1 f_2(0) + \frac{\lambda_2 \nu_1}{\nu_2} f_1(0) \tag{2.16c}$$

The resulting system of differential equations and boundary equations can be solved as in subsection 2.3.4. The solutions for the cases $\lambda_2=0$ and $\nu_1=\nu_2$ are given in appendix B. The solution method for the case $\lambda_1 \neq 0$, $\lambda_2 \neq 0$ and $\nu_1 \neq \nu_2$ is similar to the solution given in subsection 2.3.4.

2.3.7. Comparison with a discrete product model

The continuous modeling of the product flow in the insertion department was justified by two main reasons:

- The processing times of PCBs at machines were short compared with up and down times of machines.
- Discrete modeling leads to complex models which are difficult to analyze.

In this section, the first of these two arguments will be investigated. Assume there are different time scales involved in the operation of a machine, like processing times on the one hand and life and repair times on the other. If there is a great difference in magnitude between these time scales, then it is intuitively justified to approximate the product flow by a continuous one. The question is of course, at what difference this is allowed. To this end the discrete product analogue of the model introduced in subsection 2.3.2 is investigated, for the case of equal machine speeds.

For the discrete product analogue of the model of subsection 2.3.2. we follow Gershwin and Schick (1983) in the assumption that a machine that is up does not start an operation if the succeeding buffer is full. This means that no extra machine states are needed to represent blocking of machines. The machines have no capacity for products. A full list of assumptions is given below.

(1) The upstream machine is never starved, the downstream machine never blocked.

(2) The system is considered at times 0,T,2T,3T,.. . In a period (kT,(k+1)T) both machines process T products (when not under repair, starved or blocked). Both machines start processing the T products at the same instant.

(3) Machines have geometrically distributed times between failure and times to repair. If machine i is up in a period (kT,(k+1)T) there is a constant probability $p_i(T)$ that it fails during that period. Thus the mean time that machine i is up is $1/p_i(T)$ (measured in periods). If machine i is down at the beginning of a period, there is a constant probability $r_i(T)$ that it is repaired during that period. Hence, the mean repair time (in periods) of that machine is $1/r_i(T)$.

(4) Machines fail at the same rate when blocked or starved.

(5) When a machine breaks down, the T products it was operating on remain in the upstream storage. After the machine is repaired the processing resumes (hence products are not scrapped).

(6) By convention, a period (kT,(k+1)T) begins with a transition in storage level and ends with a transition in the machine operating conditions.

(7) No two transitions in the machine operating conditions occur in the same period (kT,(k+1)T).

(8) A machine starts operating only if at least T products are available in the upstream buffer and if at least T empty slots are present in the downstream buffer.

Most of the assumptions made here are standard for T=1 and are also present in the discrete product models in Table 2.2.1 (except for the papers of Gershwin and Berman (1981) and Buzacott (1972)). It is found from Table 2.2.1 that all the discrete-product models mentioned there are closely related to this model. The closest related model is that of Gershwin and Schick (1979) or Gershwin and Schick (1983) (with T=1). Different are the assumptions on the observation instant, state dependent failures and the possible occurrence of events of slight probability, such as simultaneous (that is within a single machine cycle) failures of the machines, simultaneous repairs and simultaneous failure of the one and repair of the other machine.

The discrete-product model can be analyzed in the same way as the continuous model. Instead of differential equations we obtain difference equations. These equations are (with notations as in subsection 2.3.3, K is now an integer).

$$f_1(x) = \alpha_1(x)T + (1-p_1(T)-r_2(T))f_1(x+T) + p_1(T)f_3(x+T) + r_2(T)f_4(x+T), \quad 0\leq x\leq K-T$$

$$f_2(x) = \alpha_2(x)T + (1-r_1(T)-p_2(T))f_2(x-T) + r_1(T)f_4(x-T) + p_2(T)f_3(x-T), \quad T \le x \le K$$

$$f_3(x) = \alpha_3(x)T + (1-r_1(T)-r_2(T))f_3(x) + r_1(T)f_1(x) + r_2(T)f_2(x), \quad 0 \le x \le K$$

$$f_4(x) = \alpha_4(x)T + (1-p_1(T)-p_2(T))f_4(x) + p_1(T)f_2(x) + p_2(T)f_1(x), \quad 0 \le x \le K$$

The boundary conditions are

$$f_1(K) = \alpha_1(K)T + (1-p_1(T)-r_2(T))f_1(K) + p_1(T)f_3(K) + r_2(T)f_4(K)$$

$$f_2(0) = 0$$

By letting T be real–valued and $p_i(T)$ and $q_i(T)$ be such, that $\lim_{T\downarrow 0} p_i(T)/T =: \lambda_i$ and $\lim_{T\downarrow 0} q_i(T)/T =: \mu_i$ and letting $T\downarrow 0$ we obtain the system of differential equations of Appendix A.2 with $\nu=1$. This shows that the case of the continuous product flow is the limiting situation of the discrete product flow case.

The cycle cost for the discrete–product model is calculated in appendix A.3. The analysis is similar to that in appendix A.2. It is shown that the cycle cost is close to the cycle cost in the continuous–flow model with equal machine speeds.

In order to get an idea as to how close the throughputs and mean buffer contents are, in the continuous and discrete product models, the relative difference of these quantities is investigated. It is assumed that T=1, $\nu_1=\nu_2=1$ and $p_i(T):=p_iT$, $q_i(T):=q_iT$, $p_i:=\lambda_i$, $q_i:=\mu_i$, i=1,2. In Figures 2.3.4 the relative difference for the throughput $(v_{dis}(K) - v_{con}(K))/v_{dis}(K)$ is given for varying parameters (in %). In Figures 2.3.5 the relative difference in mean buffer contents is indicated. In all cases $p_i:=\lambda_i$ and $q_i:=\mu_i$, i =1,2. From Figures 2.3.4 it can be noted that the relative differences in throughput are slight if either λ_1 or λ_2 is small. For instance, the differences are less than 8% if either $\lambda_1 < 0.09$ or $\lambda_2 < 0.09$. The relative errors in throughput are less than 5% for all buffer capacities $\ge$ 2, if both $\lambda_1<0.25$ and $\lambda_2<0.25$. Hence, for the relative differences in throughput to be less than 5%, the time scales of processing times and lifetimes have to differ by a factor of 4 only, for both machines. A similar statement can be made for the relative differences in buffer contents, although those differences are greater. Note that, even for K=10, the relative differences in mean buffer content are great for $\lambda_1>0.25$ (especially when λ_2 is small simultaneously). However, in this case the

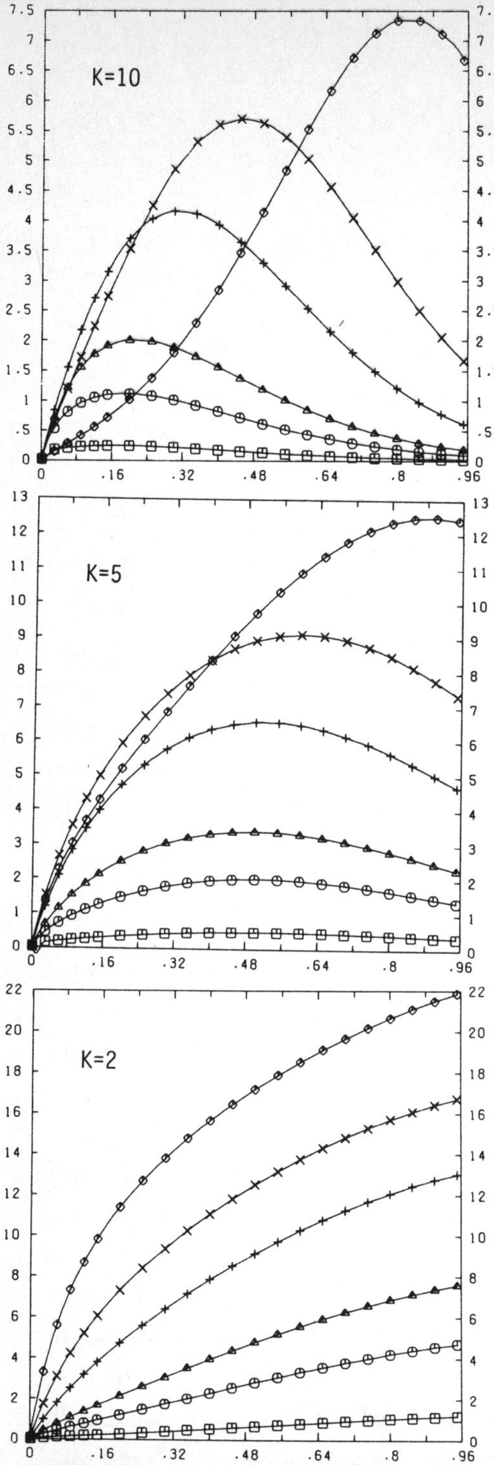

Figures 2.3.4. Relative difference in throughput

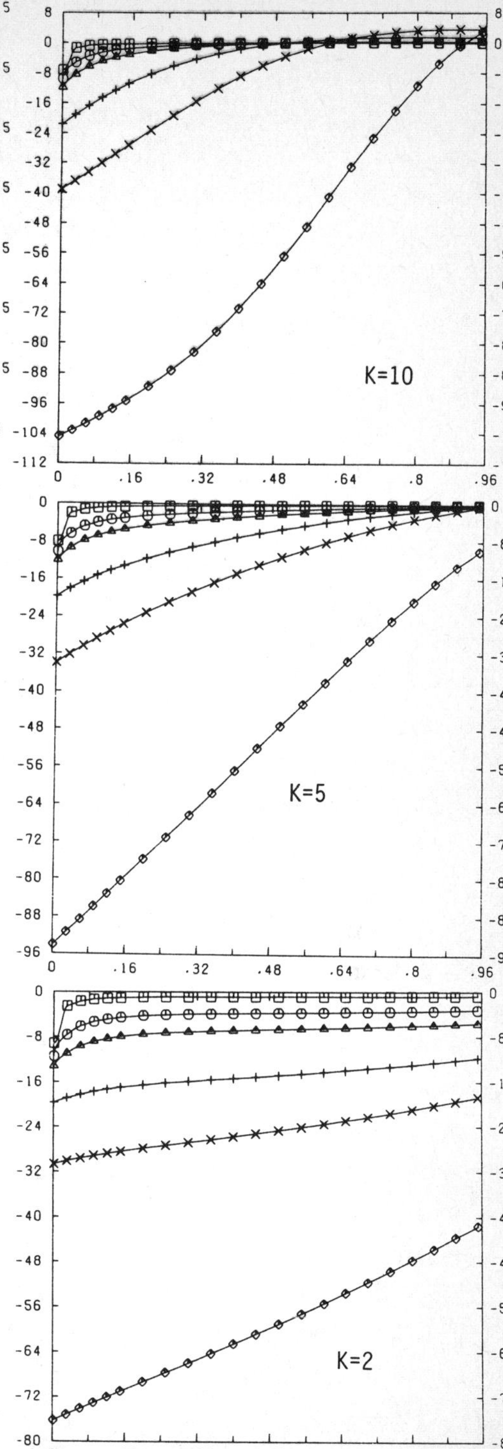

Figures 2.3.5. Relative difference in mean buffer contents.

On the horizontal axis λ_2 is shown. The buffer capacities are 10, 5 and 2, respectively. $\nu_1=\nu_2=1$, $\mu_1=0.1$, $\mu_2=0.09$; □:$\lambda_1=0.009$; o:$\lambda_1=0.045$; Δ:$\lambda_1=0.09$; +:$\lambda_1=0.25$; ×:$\lambda_1=0.45$; ◇:$\lambda_1=0.9$.

mean buffer drift σ is much smaller than zero, so that the buffer is nearly empty. Relative differences may be great, but the absolute differences in mean buffer contents are slight in this case.

2.4 MORE COMPLEX MACHINES

The assumptions made in the model of the previous section are rather restrictive. It is assumed that machines work at a constant speed and that they are susceptible to failures, with exponential life and repair times. In reality, different life and repair–time distributions may be needed, or groups of parallel machines may have to be modeled. In this section, the model will be extended, so that it can be used to model other life and repair–time distributions and groups of parallel machines on both sides of the buffer. Such groups of parallel machines were also modeled by Mitra (1986). This extended model was introduced in section 1.4. It can be used for the analysis of the insertion department of section 2.1. The model is introduced in subsection 2.4.1. Interpretations of the model are given in subsection 2.4.2. The analysis of the model is treated in subsection 2.4.3.

2.4.1 The model

In the present subsection an extension of the model of section 2.3 is introduced, which will allow the analysis of more complex machines, as will be explained later. The model of section 2.3 is extended by allowing a larger number of states per work station. The sojourn time in each machine state is exponentially distributed and the transitions from one state to another are determined by an irreducible Markov process. A Markov process is called irreducible if each state can be reached from each other state. A speed is associated with each possible machine state.

More precisely, let the production rates of machine M_i (i=1,2) be $\nu_1^i,..,\nu_{N_i}^i$(not necessarily different). These rates are given in decreasing order. M_i works at speed ν_r^i during an exponentially distributed interval (parameter λ_r^i). After a period at speed ν_r^i a transition takes place. With probability p_{rs}^i the transition is to a state having a speed ν_s^i. The matrix $\mathbf{P}^i=(p_{rs}^i)$ is an irreducible Markov matrix. It is assumed that $p_{jj}^i=0$, for i=1,2 and all j. Furthermore, it is assumed that $\nu_1^1>\nu_{N_2}^2$ and $\nu_{N_1}^1<\nu_1^2$ (otherwise the buffer is useless). The state of the system is determined by the machine states $(k,\ell) \in \{1,..,N_1\}\times\{1,..,N_2\}$, denoting the states of the first and second machine, and by the buffer level $x \in [0,K]$.

The entrance of $(N_1,1,0)$ is chosen as regeneration point. Note that this is a good choice, since $\nu^1_{N_1} < \nu^2_1$.

In order to calculate the quantities defined in Table 2.3.1 for this model, we proceed as in section 2.3.3. That is, we define cost rates in each state, costs till absorption and cycle costs. For a right choice of the cost rates in each state, the cycle costs equal the required variable. The cycle cost can be calculated by solving a system of differential equations.

Let $\mathbf{A_i}$ be the generator of the Markov process on the states of machine M_i, $i=1,2$,

$$\text{that is } \mathbf{A_i} = \begin{bmatrix} -\lambda^i_1 & p^i_{12}\lambda^i_1 & \cdots & p^i_{1N_i}\lambda^i_1 \\ p^i_{21}\lambda^i_2 & -\lambda^i_2 & \cdots & p^i_{2N_i}\lambda^i_2 \\ \vdots & \vdots & & \vdots \\ p^i_{N_i1}\lambda^i_{N_i} & p^i_{N_i2}\lambda^i_{N_i} & \cdots & -\lambda^i_{N_i} \end{bmatrix}$$

Let $\mathbf{D_i}$ be the $N_i \times N_i$ diagonal matrix $\mathrm{Diag}(\nu^i_1,..,\nu^i_{N_i})$ and let the cost rate $\alpha_{(k,\ell)}(x)$ and the cost functions $f_{(k,\ell)}(x)$ be defined as in section 2.3.3, for $(k,\ell) \in \{1,..,N_1\} \times \{1,..,N_2\}$.

It is possible to derive a cycle cost function and a system of differential equations for the $f_{(k,\ell)}$'s in the same way as in section 2.3.3. The cost function is given by

$$C_T = \frac{1}{\lambda^1_{N_1} + \lambda^2_1}\left[\alpha_{(N_1,1)}(0) + \sum_{i=1}^{N_1} p^1_{N_1 i}\lambda^1_{N_1} f_{(i,1)}(0) + \sum_{j=1}^{N_2} p^2_{1j}\lambda^2_1 f_{(N_1,j)}(0)\right] \tag{2.17}$$

The system of differential equations has dimension N_1N_2, since each state of M_1 has to be combined with each state of M_2. Let the machine states be given in increasing lexicographic order, then the system of differential equations is given by

$$-\mathbf{D}\mathbf{f}'(x) = \boldsymbol{\alpha}(x) + \mathbf{A}\mathbf{f}(x) \tag{2.18}$$

In (2.18), $\mathbf{D}$ is the *drift matrix*. It is diagonal of dimension $(N_1 \times N_2)$ with the element in entry $((k,\ell),(k,\ell))$ (lexicographically ordered) equal to $\nu^1_k - \nu^2_\ell$ $((k,\ell) \in \{1,..,N_1\} \times \{1,..,N_2\})$. $\mathbf{A}$ is the square N_1N_2–dimensional matrix $\mathbf{A_1} \otimes \mathbf{I}(N_2) + \mathbf{I}(N_1) \otimes \mathbf{A_2}$, where $\mathbf{I}(n)$ is the $n \times n$ unity matrix. The symbol $\otimes$ denotes the Kronecker product (see Bellman, 1970)[1].

Let d_0, d_+ and d_- be the number of zero, positive and negative diagonal elements of the drift matrix **D**. The dimension of the system (2.18) can be reduced by d_0, by substituting the d_0 ordinary equations with 0 on the left-hand side, in the other differential equations. There are $N_1N_2-d_0$ boundary equations, given at either x=0 or x=K, depending on the sign of the drift matrix (d_+ equations at x=K and d_- at x=0). One of the equations at x=0 is

$$f_{(N_1,1)}(0) = 0$$

The aim is again to find the throughput, mean buffer content and similar quantities as defined in Table 2.3.1. Since we do not have up–down machines, it is not possible to define the number of switches from up to down states per cycle, O_i (input) and O_o (output) (see Table 2.3.1). It will be shown in Chapter 4, that it is convenient for approximation purposes to define similar quantities for this model. It is possible for the input of the system, to distinguish a speed level z, such that the input speed of the two–stage line is sometimes greater than z and sometimes smaller. Further this speed level z can be chosen arbitrarily. Such a speed level z can also be distinguished for the output. Let O_i^z be the average number of switches in a cycle, of the input speed from a level less than or equal to z, to a level greater than z. Note that this quantity equals the average number of switches in a cycle, from an input speed greater than z, to an input speed less than, or equal to z. O_o^z is defined similarly for the output speed.

To obtain the other quantities defined in Table 2.3.1 and the quantities O_i^z and O_o^z, the corresponding values for $\alpha_{(k,\ell)}(x)$ have to be substituted in (2.17) and (2.18). The substitutions that have to be made are indicated in Table 2.4.1. This table corresponds to Table 2.3.2.

---

[1]If **G** is a square m–dimensional matrix and **H** is square n–dimensional, then **G**⊗**H** is square mn–dimensional. **G**⊗**H** = $\{G(k,\ell)\mathbf{H}\}$, which means that **G**⊗**H** is obtained from **G** by replacing element $G(k,\ell)$ by a square matrix, namely $G(k,\ell)\mathbf{H}$.

	$\alpha_{(k,\ell)}(x)$	$\alpha_{(k,\ell)}(0)$	$\alpha_{(k,\ell)}(K)$
T	1	1	1
P	ν^2_ℓ	$\min\{\nu^1_k,\nu^2_\ell\}$	ν^2_ℓ
I_ν	$\delta_{\{\nu^1_k=\nu\}}$	$\delta_{\{\nu^1_k=\nu\}}$	$\delta_{\{\nu^1_k=\nu\}}\cdot\delta_{\{\nu^1_k\leq\nu^2_\ell\}}$ + $\delta_{\{\nu^2_\ell=\nu\}}\cdot\delta_{\{\nu^1_k>\nu^2_\ell\}}$
O^z_o	$\delta_{\{(k,\ell)\in Z^1\}}\sum_{\{j\in N_2\mid(k,j)\in Z^2\}}\lambda^2_\ell p^2_{\ell j}$ + $\delta_{\{(k,\ell)\in Z^3\}}\sum_{\{j\in N_2\mid(k,j)\in Z^4\}}\lambda^2_\ell p^2_{\ell j}$	$\delta_{\{(k,\ell)\in Z^2\}}\sum_{\{i\in N_1\mid(i,\ell)\in Z^4\}}\lambda^1_k p^1_{ki}$ + $\delta_{\{(k,\ell)\in Z^3\}}\sum_{\{j\in N_2\mid(k,j)\in Z^4\}}\lambda^2_\ell p^2_{\ell j}$	$\delta_{\{(k,\ell)\in Z^1\}}\sum_{\{j\in N_2\mid(k,j)\in Z^2\}}\lambda^2_\ell p^2_{\ell j}$ + $\delta_{\{(k,\ell)\in Z^3\}}\sum_{\{j\in N_2\mid(k,j)\in Z^4\}}\lambda^2_\ell p^2_{\ell j}$
O^z_i	$\delta_{\{(k,\ell)\in Z^1\}}\sum_{\{i\in N_1\mid(i,\ell)\in Z^3\}}\lambda^1_k p^1_{ki}$ + $\delta_{\{(k,\ell)\in Z^2\}}\sum_{\{i\in N_1\mid(i,\ell)\in Z^4\}}\lambda^1_k p^1_{ki}$	$\delta_{\{(k,\ell)\in Z^1\}}\sum_{\{i\in N_1\mid(i,\ell)\in Z^3\}}\lambda^1_k p^1_{ki}$ + $\delta_{\{(k,\ell)\in Z^2\}}\sum_{\{i\in N_1\mid(i,\ell)\in Z^4\}}\lambda^1_k p^1_{ki}$	$\delta_{\{(k,\ell)\in Z^3\}}\sum_{\{j\in N_2\mid(k,j)\in Z^4\}}\lambda^2_\ell p^2_{\ell j}$ + $\delta_{\{(k,\ell)\in Z^2\}}\sum_{\{i\in N_1\mid(i,\ell)\in Z^4\}}\lambda^1_k p^1_{ki}$
S_ν	$\delta_{\{\nu^2_\ell=\nu\}}$	$\delta_{\{\nu^2_\ell=\nu\}}\cdot\delta_{\{\nu^1_k\geq\nu^2_\ell\}}$ + $\delta_{\{\nu^1_k=\nu\}}\cdot\delta_{\{\nu^1_k<\nu^2_\ell\}}$	$\delta_{\{\nu^2_\ell=\nu\}}$
Q_1	0	0	$\delta_{\{\nu^2_\ell=0\}}$
Q_2	0	$\delta_{\{\nu^1_k=0\}}$	0
Y	x	0	K
PL	$\delta_{\{x\geq L\}}$	$\delta_{\{0\geq L\}}$	1

Table 2.4.1. Determination of some line quantities (see Table 2.3.1 for definitions).
$Z^1 = \{(k,\ell) \in N^1\times N^2 \mid \nu^1_k \leq z,\ \nu^2_\ell \leq z\}$; $Z^2 = \{(k,\ell) \in N^1\times N^2 \mid \nu^1_k \leq z < \nu^2_\ell\}$
$Z^3 = \{(k,\ell) \in N^1\times N^2 \mid \nu^2_\ell \leq z < \nu^1_k\}$; $Z^4 = \{(k,\ell) \in N^1\times N^2 \mid \nu^1_k > z,\ \nu^2_\ell > z\}$.

2.4.2 Interpretations of the model

In this section various possibilities will be given to use the model of the previous section.

In the first place it is possible, to represent arbitrary phase-type distributions for the time that a machine is running at a certain speed. This can be realized by choosing some of the ν_j^i's equal (i=1,2), an appropriate matrix $\mathbf{P}^i$ and rates λ_j^i.

In the second place it is possible to model a group of parallel machines. We have already referred to this at the beginning of section 2.4. Assume that upstream of the buffer we have n parallel machines, with a speed ν_i, failure rate λ_i and repair rate μ_i, for machine i. Then the machine states are given by an n-tuple $(\delta_1,...,\delta_n)$, where δ_j is shorthand for 1, if machine j is up and 0, if machine j is down. The production rate of the aggregate machine representing the n parallel machines, corresponding to the state $(\delta_1,...,\delta_n)$ is $\sum_{j=1}^{n} \nu_j \delta_j$. The duration of an interval with this production rate is exponentially distributed with parameter $\sum_{j=1}^{n} [\lambda_j \delta_j + \mu_j(1-\delta_j)]$. If the states of the aggregate machine are ordered in decreasing lexicographic order, then the element p_{12}, for instance, in the transition matrix $\mathbf{P} = (p_{ij})$ between these states equals $\frac{\lambda_n}{\lambda_1+..+\lambda_n}$. It is easy to see that $\mathbf{P}$ is irreducible. Note that the model is more general than that of Mitra (1986), who only studies identical parallel machines. In the case of *identical* parallel machines, the model simplifies, since then only N+1 machine states are needed to represent N parallel up-down machines. Such a machine state corresponds to the number of up-machines at that moment. Note that, in the insertion department described in section 2.1, we have to do with identical parallel machines.

In the third place, this model could be used for aggregate modeling purposes. Assume that we want to model in an aggregate way a two-stage line which produces many products. We take it that each product has its specific production rate at each machine M_i. Between the subsequent production runs of any two of these products, there is an exponential switch-over time. At aggregate level, these products cannot be distinguished from each other. However, fluctuations in the machine speeds, measured in product units per time unit, can be distinguished. This could be modeled by letting machine M_i produce at rates $(\nu_1^i,..,\nu_{N_i}^i,\nu_{N_i+1}^i)$, where rate ν_j^i corresponds to production of product j. Rate $\nu_{N_i+1}^i$ equals 0. It represents switch-overs (and possibly machine failures). The square N_i+1-dimensional transition matrices $\mathbf{P}^i$ represent set-ups for different products, that is, $p^i_{rs}=0$ for all $r\neq N_i+1$ and $s\neq N_i+1$, $p^i_{r,N_i+1}=1$, $\sum_{s=1}^{N_i} p^i_{N_i+1,s}=1$. $p^i_{N_i+1,s}$ may be chosen equal to the average frequency with which product s is set up. Note however, that the downstream machine is not really able to distinguish the products in the buffer. This kind of aggregate modeling is applied in Chapter 7.

2.4.3 Model analysis

For the acquisition of the quantities in Table 2.4.1, we have to calculate the cycle cost (2.17), with the appropriate substitution for the cost rates $\alpha_{(k,\ell)}(x)$. To this end (2.18) has to be solved and the solution for $\mathbf{f}(0)$ has to be substituted in (2.17). To obtain the solution of (2.18) we again have to find a homogeneous and a particular solution. For the determination of the homogeneous solution, we can use, as in subsection 2.3.4, the analysis of Mitra (1986), to prove that the eigenvalues of **A** are real and that the number of positive, zero and negative eigenvalues depend on the *mean drift* $\sigma := \text{eff}_1 - \text{eff}_2$ and on the signs of the entries of **D**.

The *efficiency* of M_i is defined as $\text{eff}_i := (\mathbf{D}_i\mathbf{w}_i)^T\cdot\mathbf{1}_{N_i}$, where $\mathbf{1}_n$ is the n–vector consisting of ones and $\mathbf{w}_i$ is the equilibrium vector corresponding to $\mathbf{A}_i$ (that is, given by $\mathbf{w}_i^T\mathbf{A}_i = \mathbf{0}$ and $\mathbf{w}_i^T\cdot\mathbf{1}_{N_i} = 1$). Note that this definition is consistent with the definition given in section 2.3.4.

If the $(N_1\times N_2)$–dimensional vector $\mathbf{w}$ is defined as $\mathbf{w} = \mathbf{w}_1 \otimes \mathbf{w}_2$, then it can be proven easily that

$$\mathbf{w}^T\mathbf{A} = \mathbf{0},\ \mathbf{w}^T\cdot\mathbf{1}_{N_1N_2} = 1 \text{ and } \sigma = (\mathbf{D}\mathbf{w})^T\cdot\mathbf{1}_{N_1N_2} \tag{2.19}$$

The following division of the eigenvalues of **A** follows from theorem 1 in Mitra (1986).

	# positive eigenvalues	multiplicity of eigenvalues at 0	# negative eigenvalues
$\sigma < 0$	d_+	1	$d_- - 1$
$\sigma = 0$	$d_+ - 1$	2	$d_- - 1$
$\sigma > 0$	$d_+ - 1$	1	d_-

Table 2.4.2. Sign division of eigenvalues of A.

It is straightforward to solve the system of differential equations (2.18). The homogeneous and a particular solution can be found similar to subsection 2.3.4. However, computational problems may arise when $N_1N_2-d_0$ is large (especially in the calculation of eigenvalues, eigenvectors and the constants in the homogeneous solution). An approximation method to overcome these problems is given in section 2.7.

The next section (2.5) is not necessary for the analysis of the insertion department. It deals with more general buffer control policies in two–stage lines.

2.5 CONTROL OF THE TWO–STAGE LINE

The analysis in section 2.3 refers to machine lines with *physical* buffer constraints. Our main objective however is to analyze the effect of capacity control rules (also in logistic lines). Logistic lines differ in the following aspects from machine lines.

1. In logistic lines, the last "machine" is demand. In a machine line as sketched in Fig. 2.3.1, M_2 stops (or is slowed down) as soon as the preceding buffer gets empty. Demand may be *backlogged*, however. This may be modeled by a negative lower bound for the allowed buffer contents. The buffer range can be modeled then as $[\kappa,K]$, with $\kappa<0$. In case backlogging is *complete*, $\kappa=-\infty$, otherwise backlogging is *restricted.*
2. The maximum buffer content is often not really hard, although it can become hard by the applied control. If M_2 is faster on average than M_1, or $eff_2>eff_1$, then it is not necessary to control the buffer content. Such situations can be modeled by letting $K=\infty$.
3. In logistic lines the work stations may be controlled by $<m,M>$ switching policies (see also section 1.3). A work station is allowed to keep producing as long as the buffer content remains below $M>m$. As soon as the buffer content increases to M, the work station stops. It is allowed to start operating again as soon as the buffer content decreases to the level m. Such policies are used to restrict the number of up–down switches of M_1.

In section 2.5.1 two–stage lines, with buffer contents bounded from below by κ and from above by K, with $-\infty\leq\kappa<K\leq\infty$, are studied. Switching policies of $<m,M>$ type are studied in section 2.5.2.

2.5.1 Infinite buffers and backlogging

In this subsection buffers with range $[\kappa,K]$ are studied. The following definition is convenient.

Two two-stage lines with buffer ranges [a,b] and [a+C,b+C], respectively ($a<b$, $C\in\mathbb{R}_+$), are called *buffer-shifted equivalent* if they have the same throughput and if the average buffer content of the second line equals that of the first line, plus C.

Note that two identical lines, but with buffer ranges [a,b] and [a+C,b+C] and the same slowing down behavior at both the lower and upper buffer bound are buffer-shifted equivalent.

The following cases can be distinguished.

1. The case $-\infty<\kappa<K<\infty$.

If $\kappa<0$, then only restricted backlogging is possible. If $\kappa\geq0$, then we have to do with *lost sales.* The solution is similar to subsection 2.3.4. The system is buffer-shifted equivalent to an identical system, but with buffer range $[0,K-\kappa]$.

2. The case $-\infty<\kappa$, $K=\infty$.

In this case we have to distinguish between *stable* and *unstable* systems. If $-\infty<\kappa$, $K=\infty$, then the system is called *stable* if $\sigma<0$. Recall that $\sigma = \text{eff}_1-\text{eff}_2$, the mean drift of the system. The system is called *unstable* if $\sigma\geq0$.

In the unstable case no steady state behavior will be reached. Therefore we will restrict ourselves to the stable case. In the stable case, the line is a "pull system", which ensures that the buffer contents do not grow to infinity.

Boundary condition (2.2a) with x=K, does not hold any more. If $\nu_1>\nu_2$, then also boundary condition (2.5a) does not hold any more. Hence we are short of either one ($\nu_1\leq\nu_2$) or two boundary conditions ($\nu_1>\nu_2$). The missing boundary conditions are obtained by using the sign division of the eigenvalues, indicated in Table 2.3.3.

If $\nu_1\leq\nu_2$, then there is, according to Table 2.3.3, exactly one eigenvalue, say ρ_3 (with eigenvector $\mathbf{e}_3$) which is positive (notation of section 2.3.4, see formula (2.9)). Since the solution of the homogeneous equation (2.9) would grow exponentially fast if $C_3\neq0$ (recall that $K=\infty$), the stability of the system implies that the boundary condition (2.2a) has to be replaced by the condition $C_3=0$. This leads again to sufficiently many boundary conditions, the system can be solved as before.

If $\nu_1>\nu_2$, then **A** (see 2.7b) has two positive eigenvalues, say ρ_2 and ρ_3, with eigenvectors $\mathbf{e_2}$ and $\mathbf{e_3}$ (see formula (2.9)). Therefore boundary equations (2.2a) and (2.5a) have to be replaced by $C_2=0$ and $C_3=0$. The remaining analysis is similar to the analysis in section 2.3.4.

From the results in appendix A.1 and A.2, it is possible to obtain explicit limiting values for the throughput and mean buffer content for the cases $\lambda_2=0$ and $\nu_1=\nu_2$, respectively, by letting $K \to \infty$.

3. The case $\kappa=-\infty$, $K<\infty$.
In this case backlogging is complete. The system is called *stable* if $\sigma>0$, and *unstable* if $\sigma\leq 0$. It is again assumed that the system is stable, which means that we have to do with a "push system". The analysis is similar to the case $K = \infty$, apart from the fact that another regeneration point has to be chosen, for instance (1,K).

4. The case $\kappa=-\infty$, $K=\infty$.
In this case the system is unstable for all parameter settings.

2.5.2. Control by $<m,M>$ switching policies

Until now it was assumed that M_1 is not blocked or slowed down as long as the buffer is not full. Other operating policies for M_1 may be attractive however. M_1 may for instance be controlled by an $<m,M>$ switching policy. It is assumed that the lower bound on the buffer content is κ, with $\kappa\in\mathbb{R}$ and $\kappa<m$. Note that the control in subsection 2.3.2 can also be interpreted as a special switching policy with $m=M$.

For related literature on switching policies in continuous-flow lines, see De Kok (1987) and Hopp (1988). For related work in discrete-product lines, see Altiok (1986a) and Buzacott (1982).

Here the two-stage system with unreliable machines, operating with this policy, is analyzed. The analysis is along the same lines as in section 2.3. We distinguish regeneration points, cost rates in each state and cost functions till absorption. The cycle cost can be obtained by solving a system of differential equations in the cost functions. In the analysis we have to distinguish between

situation (a): M_1 is allowed to operate (although it may be failed), since the level M is not yet reached (from below) and
situation (b): M_1 is not allowed to operate, since the level m is not yet reached (from above).

Hence, in total there are 8 machine states as indicated in Table 2.5.1.

	1 M_1 up M_2 down	2 M_1 down M_2 up	3 M_1 down M_2 down	4 M_1 up M_2 up
a: The buffer content has not yet *increased* to M. M_1 may operate.	(1,a)	(2,a)	(3,a)	(4,a)
b: The buffer content has not yet *decreased* to m. M_1 may not operate.	(1,b)	(2,b)	(3,b)	(4,b)

Table 2.5.1. The 8 possible machine states

To obtain the system states, these 8 machine states have to be combined with the buffer level x. Hence the system state is determined by (i,j,x), with i=1,..,4; j=a,b and $x \in [\kappa, M]$. Note that if $x<m$, then situation b cannot occur any more: the system is in situation a.

The regeneration point is chosen to be an entrance to $(2,a,\kappa)$. If $\lambda_1 \neq 0$ the state will certainly return to this regeneration point.

The cost rates and cost functions are defined as follows, for i=1,..,4

$\alpha_i(x)$ is defined as the cost rate in machine state (i,a), with inventory level $x \in [\kappa, K]$

$\beta_i(x)$ is defined as the cost rate in machine state (i,b), with inventory level $x \in [\kappa, K]$.

$f_i(x)$ is the expected cost till the end of the cycle (the first entrance of $(2,a,\kappa)$), if we are now in machine state (i,a), with inventory level x. $f_2(\kappa):=0$

$g_i(x)$ is the expected cost till the end of the cycle, if we are now in machine state (i,b), with inventory level x.

The expected cost per cycle is obtained from (2.1) by replacing 0 by κ.

$$C_T = \frac{1}{\lambda_2+\mu_1}[\alpha_2(\kappa) + \lambda_2 f_3(\kappa) + \mu_1 f_4(\kappa)] \tag{2.20}$$

The system of differential equations is

$$-\nu_1 f_1'(x) = \alpha_1(x) - (\lambda_1+\mu_2)f_1(x) + \lambda_1 f_3(x) + \mu_2 f_4(x)\ , \qquad x\in[\kappa,M) \tag{2.21}$$

$$\nu_2 f_2'(x) = \alpha_2(x) - (\mu_1+\lambda_2)f_2(x) + \lambda_2 f_3(x) + \mu_1 f_4(x)\ , \qquad x\in(\kappa,M] \tag{2.22}$$

$$0 = \alpha_3(x) - (\mu_1+\mu_2)f_3(x) + \mu_1 f_1(x) + \mu_2 f_2(x)\ , \qquad x\in[\kappa,M] \tag{2.23}$$

$$(\nu_2-\nu_1)f_4'(x) = \alpha_4(x) - (\lambda_1+\lambda_2)f_4(x) + \lambda_2 f_1(x) + \lambda_1 f_2(x)\ , \qquad x\in(\kappa,M) \tag{2.24}$$

$$0 = \beta_1(x) - (\lambda_1+\mu_2)g_1(x) + \lambda_1 g_3(x) + \mu_2 g_4(x)\ , \qquad x\in[m,M] \tag{2.25}$$

$$\nu_2 g_2'(x) = \beta_2(x) - (\mu_1+\lambda_2)g_2(x) + \lambda_2 g_3(x) + \mu_1 g_4(x)\ , \qquad x\in(m,M] \tag{2.26}$$

$$0 = \beta_3(x) - (\mu_1+\mu_2)g_3(x) + \mu_1 g_1(x) + \mu_2 g_2(x)\ , \qquad x\in[m,M] \tag{2.27}$$

$$\nu_2 g_4'(x) = \beta_4(x) - (\lambda_1+\lambda_2)g_4(x) + \lambda_2 g_1(x) + \lambda_1 g_2(x)\ , \qquad x\in(m,M] \tag{2.28}$$

The boundary conditions are

$$f_1(M) = g_1(M) \tag{2.21a}$$

$$f_2(\kappa) = 0 \tag{2.22a}$$

$$f_4(M) = g_4(M)\ , \qquad \text{if } \nu_1>\nu_2 \tag{2.24a}$$

$$0 = \alpha_4(\kappa) - (\lambda_1+\lambda_2)f_4(\kappa) + \lambda_2 f_1(\kappa) + \lambda_1 f_2(\kappa)\ , \qquad \text{if } \nu_1\le\nu_2 \tag{2.24b}$$

$$g_2(m) = f_2(m) \tag{2.26a}$$

$$g_4(m) = f_4(m) \tag{2.28a}$$

The subsystem (2.25)–(2.28) can be solved separately, by using boundary equations (2.26a) and (2.28a). This can be done by substituting $g_1(x)$, obtained from (2.25), and $g_3(x)$, obtained from (2.27), in (2.26) and (2.28). This results in the system

$$\nu_2 g_2'(x) = \beta_2^*(x) - \frac{\mu_1(\lambda_1+\lambda_2+\mu_1+\mu_2)}{(\lambda_1+\mu_1+\mu_2)}(g_2(x) - g_4(x)) \tag{2.29}$$

$$\nu_2 g_4'(x) = \beta_4^*(x) + \frac{\lambda_1(\lambda_1+\lambda_2+\mu_1+\mu_2)}{(\lambda_1+\mu_1+\mu_2)}(g_2(x) - g_4(x)), \tag{2.30}$$

where $\beta_2^*(x) = \beta_2(x) + \dfrac{\lambda_2\mu_1\beta_1(x) + \lambda_2(\lambda_1+\mu_2)\beta_3(x)}{\mu_2(\lambda_1+\mu_1+\mu_2)}$

and $\beta_4^*(x) = \beta_4(x) + \dfrac{\lambda_2(\mu_1+\mu_2)\beta_1(x) + \lambda_1\lambda_2\beta_3(x)}{\mu_2(\lambda_1+\mu_1+\mu_2)}$

To obtain the mean cycle length T or the mean production per cycle P, the corresponding values for $\alpha_i(x)$ and $\beta_i(x):=\alpha_i(x)$, indicated in Table 2.3.2, with 0 replaced by κ, have to be substituted in equations (2.21)–(2.28a). For these quantities, $\alpha_i(x) = \alpha_i$ and $\beta_i(x) = \beta_i$ and hence $\overset{*}{\beta}_i(x) = \overset{*}{\beta}_i$, for $x\in(\kappa,M)$, independent of x. The solution of (2.29)–(2.30), with boundary equations (2.26a), (2.28a) is then, for $x\in[m,M]$:

$$g_2(x) = \left[\frac{\overset{*}{\beta}_2}{\nu_2} - \frac{\mu_1(\overset{*}{\beta}_2-\overset{*}{\beta}_4)}{\nu_2(\mu_1+\lambda_1)}\right](x-m) + \frac{\mu_1}{\lambda_1+\mu_1}\left[f_2(m)-f_4(m)-\frac{(\overset{*}{\beta}_2-\overset{*}{\beta}_4)}{\nu_2\rho}\right](e^{-\rho(x-m)}-1) + f_2(m), \qquad (2.31)$$

$$g_4(x) = \left[\frac{\overset{*}{\beta}_4}{\nu_2} + \frac{\lambda_1(\overset{*}{\beta}_2-\overset{*}{\beta}_4)}{\nu_2(\mu_1+\lambda_1)}\right](x-m) - \frac{\lambda_1}{\lambda_1+\mu_1}\left[f_2(m)-f_4(m)-\frac{(\overset{*}{\beta}_2-\overset{*}{\beta}_4)}{\nu_2\rho}\right](e^{-\rho(x-m)}-1) + f_4(m), \qquad (2.32)$$

where $\rho := \dfrac{(\lambda_1+\mu_1)(\lambda_1+\lambda_2+\mu_1+\mu_2)}{\nu_2(\lambda_1+\mu_1+\mu_2)}$

The boundary equations (2.21a)–(2.24b) corresponding to the system (2.21)–(2.24) can be rewritten, by using (2.25), (2.27), (2.31) and (2.32), as

$$\begin{aligned} f_1(M) = {} & \frac{(\mu_1+\mu_2)\beta_1(M)+\lambda_1\beta_3(M)}{\mu_2(\lambda_1+\mu_1+\mu_2)} + \frac{\lambda_1\mu_2}{\mu_2(\lambda_1+\mu_1+\mu_2)}\left[\left[\frac{\overset{*}{\beta}_2}{\nu_2} - \frac{\mu_1(\overset{*}{\beta}_2-\overset{*}{\beta}_4)}{\nu_2(\lambda_1+\mu_1)}\right](M-m) + \right. \\ & \left. + \frac{\mu_1}{\lambda_1+\mu_1}\left(f_2(m) - f_4(m) - \frac{\overset{*}{\beta}_2-\overset{*}{\beta}_4}{\nu_2\rho}\right)(e^{-\rho(M-m)}-1) + f_2(m)\right] + \\ & + \frac{\mu_2(\mu_1+\mu_2)}{\mu_2(\lambda_1+\mu_1+\mu_2)}\left[\left[\frac{\overset{*}{\beta}_4}{\nu_2} + \frac{\lambda(\overset{*}{\beta}_2-\overset{*}{\beta}_4)}{\nu_2(\lambda_1+\mu_1)}\right](M-m) + \right. \\ & \left. - \frac{\lambda_1}{\lambda_1+\mu_1}\left(f_2(m) - f_4(m) - \frac{\overset{*}{\beta}_2-\overset{*}{\beta}_4}{\nu_2\rho}\right)(e^{-\rho(M-m)}-1) + f_4(m)\right] \end{aligned} \qquad (2.33)$$

$$f_2(\kappa) = 0 \qquad (2.34)$$

$$f_4(M) = \left[\frac{\overset{*}{\beta}_4}{\nu_2} + \frac{\lambda_1(\overset{*}{\beta}_2-\overset{*}{\beta}_4)}{\nu_2(\lambda_1+\mu_1)}\right](M-m) - \frac{\lambda_1}{\lambda_1+\mu_1}\left[f_2(m)-f_4(m)-\frac{(\overset{*}{\beta}_2-\overset{*}{\beta}_4)}{\nu_2\rho}\right](e^{-\rho(M-m)}-1) + f_4(m), \quad \text{if } \nu_1>\nu_2 \qquad (2.35a)$$

$$0 = \alpha_4(\kappa) - (\lambda_1+\lambda_2)f_4(\kappa) + \lambda_2 f_1(\kappa) + \lambda_1 f_2(\kappa), \quad \text{if } \nu_1\leq\nu_2 \qquad (2.35b)$$

The system (2.21)–(2.24), with boundary conditions (2.33)–(2.35b) can be solved iteratively. An initial guess can be made for $f_2(m)$ and $f_4(m)$, say $f_2^{(1)}(m)$ and $f_4^{(1)}(m)$. These can be substituted in (2.33)–(2.35b), after which the system can be

solved as before. The obtained values for $f_2(m)$ and $f_4(m)$ can be compared with $f_2^{(1)}(m)$ and $f_4^{(1)}(m)$. If the absolute difference is smaller than some prefixed number ϵ, then the procedure is finished. If not, then $f_2^{(2)}(m) := f_2(m)$, $f_4^{(2)}(m) := f_4(m)$ and the procedure is repeated, until sufficient convergence is reached. Convergence could not be proved, but in all investigated cases (see Fig. 2.5.1) convergence was reached fast.
On substituting the obtained values for $f_3(\kappa)$ and $f_4(\kappa)$ in (2.20), the cost per cycle can be obtained and hence the mean production per cycle and the mean cycle length.

To obtain Y (see Table 2.3.1) we can proceed similarly by substituting $\alpha_i(x) := x$ and $\beta_i(x) := x$, for all i and x.
The boundary equations, for $\nu_1 > \nu_2$, are (2.34) and

$$f_1(M) = \frac{M}{\mu_2} + \frac{\lambda_1\mu_2}{\mu_2(\lambda_1+\mu_1+\mu_2)}\Big[f_2(m) + \frac{\tilde{\beta}}{2\nu_2}(M^2-m^2) + \frac{\mu_1}{\lambda_1+\mu_1}\Big[f_2(m)-f_4(m)\Big]\cdot (e^{-\rho(M-m)}-1)\Big] + \frac{\mu_2(\mu_1+\mu_2)}{\mu_2(\lambda_1+\mu_1+\mu_2)}\Big[f_4(m) + \frac{\tilde{\beta}}{2\nu_2}(M^2-m^2) + - \frac{\lambda_1}{\lambda_1+\mu_1}\Big[f_2(m)-f_4(m)\Big](e^{-\rho(M-m)}-1)\Big] \tag{2.36}$$

$$f_4(M) = f_4(m) + \frac{\tilde{\beta}}{2\nu_2}(M^2-m^2) - \frac{\lambda_1}{\lambda_1+\mu_1}\Big[f_2(m) - f_4(m)\Big](e^{-\rho(M-m)}-1) \tag{2.37}$$

If $\nu_1 \leq \nu_2$, then (2.37) has to be replaced by (2.35b).

In (2.36) and (2.37),

$$\tilde{\beta} := \frac{\lambda_2+\mu_2}{\mu_2}$$

In Fig. 2.5.1 the throughput is given as a function of the buffer capacity $K=M$ for four different lines. For all lines $\kappa=0$ and $m=1.5$. The parameters of M_1 are $\nu_1=1$, $\lambda_1=0.01$ and $\mu_1=0.09$. The parameter settings for M_2 are the four cases c, listed in Table 2.3.5. Note that the throughput is in all cases considerably lower than in Figs. 2.3.2.

It is also possible to calculate the buffer content distribution, similarly as before. A cost function involving inventory holding and stock out costs (if $\kappa<0$) can be obtained by combining these costs with the buffer content distribution function.

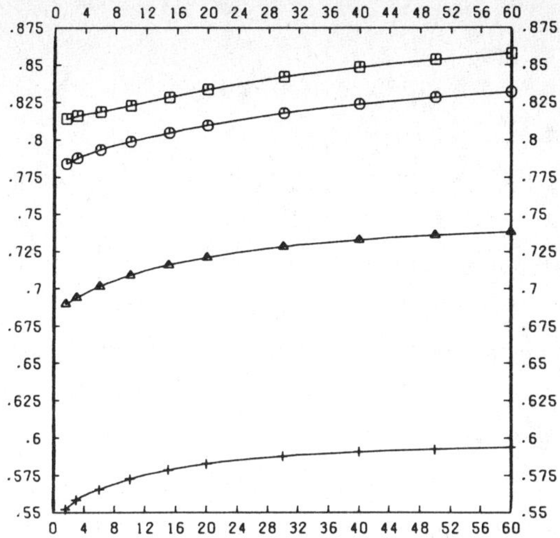

Figure 2.5.1. Throughput of a line controlled with a <m,M> switching policy as a function of M. m=1.5, ν_1=1, λ_1=0.01, μ_1=0.09, λ_2=0.01, μ_2=0.09. □:ν_2=0.99; o:ν_2=0.95; Δ:ν_2=0.833, +:ν_2=0.667.

2.6 ANALYSIS OF BUFFER BEHAVIOR

2.6.1. Buffer fluctuations

In principle we have developed all the tools needed to analyze the insertion department case of section 2.1. The analysis of section 2.4 can be applied straightforwardly to the insertion department model. However, it will be clear that incorporating all seven machines makes the state space large, which has an influence on the computation times. Therefore we will derive certain insensitivities in the model of section 2.4. To this end, we study in this subsection the behavior of the buffer content. In section 2.7 the results of this section are exploited in deriving insensitivity results which can be used for approximating complex two–stage lines by simple lines. In this section it is assumed that the liferates in the different states of the upstream and downstream machine, modeled as in section 2.4, are time–dependent. This means that they are the same when machines are forced to slow down.

The most important property of a buffer between two production capacities is that its content is allowed to fluctuate. If a buffer does not fluctuate, it has no function and can be removed from the system without loss of throughput. The buffer content behaves as a random-walk like process with barriers. The throughput of the line for the case of a finite buffer is determined by the characteristics of the random walk for the case of no bounds. If we can approximate the behavior of the buffer contents in a complex line by that of the buffer contents in a simple line, then the throughputs of these lines will approximately be the same. For such an approximation, it is sufficient to characterize the behavior of the buffer contents in case of no bounds. If the unbounded buffer content behavior in the two lines has approximately the same characteristics, then the bounded buffer content behavior will also be approximately the same, provided the lower and upper bounds of the buffer are the same. Hence also the throughput, average buffer content and all other characteristics are approximately the same. The advantage in considering unbounded buffers is that the state space can largely be reduced, since the buffer state does not have to be incorporated any longer, so that the analysis is simplified.

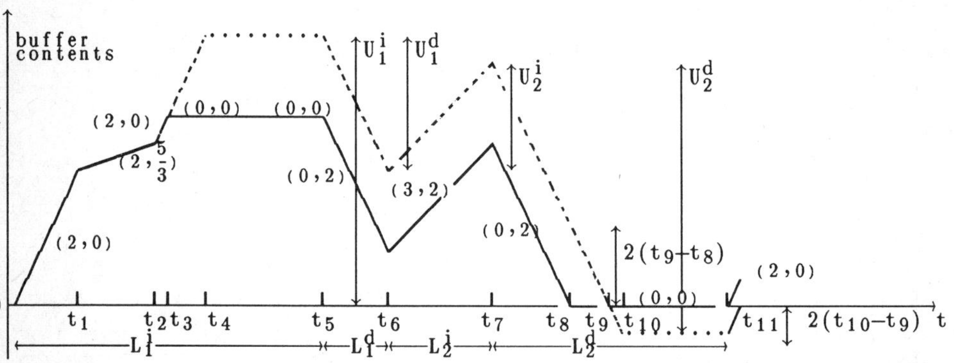

Figure 2.6.1. Behavior of the buffer contents for a certain two–stage line, with buffer capacity K and with unbounded buffer (in ...). A pair (a,b) denotes that the upstream machine runs at rate a and the downstream machine at rate b. The upstream machine can work at rates 0, 2 and 3, the downstream machine at rates 0, 5/3 and 2.
$v(\infty) = ((5/3)(t_2-t_1) + 2(t_{10}-t_5))/t_{11}$; $v(K) = ((5/3)(t_2-t_1) + 2(t_8-t_5))/t_{11}$.
$(v(\infty)-v(K))(L_1^i+L_2^i+L_1^d+L_2^d) = 2(t_{10}-t_8) = 2(t_{10}-t_9)+2(t_9-t_8) = 2(t_{10}-t_9)+2(t_4-t_3)$.

The following behavior characteristics of the unbounded buffer content are needed for the approximation. In the definitions, only unbounded stocks are considered.

L^i *Length of stock–increase period.* The time that the buffer content is non–decreasing, measured from the point where it starts increasing until it starts decreasing.

Ld *Length of stock–decrease period. The time that the buffer content is non–increasing, measured from the point where it starts decreasing until it starts increasing.*

U^i *Stock increase. The increase (in quantity) of the buffer content, measured during an increasing period of length L^i.*

U^d *Stock decrease. The decrease of the buffer content, measured during a decreasing period of length L^d.*

The probability distributions of U^i, U^d, L^i and L^d are determined by the machine states only. For a particular two–stage line, with buffer capacity K, realizations of these variables are indicated in Fig. 2.6.1. In this figure, a realization of the buffer contents is sketched as a function of time. A realization of the unbounded buffer contents (in ··) is shown as well.

Assume we have a complex two–stage line as discussed in section 2.4, with many machine states, for which calculation of the throughput and average buffer content is computationally hard. This line is to be approximated by a line, for which calculation of performance characteristics is simple. For the complex line it is computationally much simpler to calculate the first few moments of U^i and U^d, since the buffer state need not be incorporated in this calculation.

Define now the *production loss* in a line as the difference between the production with a bounded and with an unbounded buffer. Let C(K) denote the average production loss per increase/decrease step in a line with buffer capacity K. Then the *throughput loss* in this line (that is the difference in average production per unit of time between the situation with an unbounded and with a bounded buffer) times the expected time duration of an increase/ decrease step will be equal to C(K). Or, in terms of a formula,

$$(v(\infty) - v(K))(EL^i + EL^d) = C(K), \tag{2.38}$$

where v(K) is the throughput of the line with buffer capacity K and E is the expectation operator. The quantity C(K) in (2.38) is determined by the moments of U^i and U^d and the buffer capacity K (see below). It is now possible to take

another two-stage line, with the same buffer capacity K, whose throughput is easy to calculate, with the same first few moments of U^i and U^d as in the complex line. This means that $C_{ap}(K)$ of this line is approximately equal to C(K). Since it is also a straightforward matter to calculate EL^i+EL^d in both lines, it is now possible, by using (2.38) and the known throughput in the simple line, to estimate the throughput in the complex line.
This approximation is treated in section 2.7. The results in that section are based on De Koster (1987a).

In case *all* moments of U^i and U^d are equal in two lines, line a and line b, say, then $C_a(K)$ will be equal to $C_b(K)$ (assuming time-dependent rates). This can be understood, by considering two equal realizations of U^i's and U^d's in the two lines, starting with equally filled buffers of the same capacity. Assume that the buffer in line a becomes full, for the first time during the j-th stock increase period. Since the buffer capacities are equal and the U^i's and U^d's are equal in the two lines, the buffer in line b also becomes full during the j-th stock increase period. This implies that the loss in input, due to the buffer becoming full, will be the same in the two lines. The loss in output is also the same in the two lines, for similar reasons. Hence the loss in production per increase/decrease step will also be the same in the two lines, or $C_a(K)=C_b(K)$. Mention that the buffers being full (or empty) do not influence the machine states in the two lines, by the assumption of time-dependent rates. Hence also the stock increase and decrease moments are not influenced.

Note that the assumption of time-dependent rates is essential. For state-dependent rates, the quality of an approximation as indicated above will depend on the kind of state-dependence and furthermore on the differences in machine speeds.

In subsection 2.6.2 it will be demonstrated how the stock increase and decrease moments EU^i, EU^d, $E(U^i)^2$, $E(U^d)^2$, and furthermore EL^i and EL^d can be calculated for the general model of the two-stage line presented in section 2.4. Only the first two moments of U^i and U^d are calculated, but calculation of third and higher moments is similar. This subsection is (partly) based on De Koster and Wijngaard (1985).

2.6.2 Determination of the stock increase and stock decrease moments

It is assumed that the machines of the two-stage line have a structure as in section 2.4. That is, machine i has possible speeds $\nu_1,..,\nu_{N_1}$, with (exponential) rates $\lambda_1,..,\lambda_{N_1}$ and an irreducible, Markov transition matrix **P**. The speeds of machine 2 are denoted by $\varphi_1,..,\varphi_{N_2}$, the exponential rates by $\mu_1,..,\mu_{N_2}$, the transition matrix by **Q**. The speeds of both machines are ordered in a nonincreasing sequence. The machine system can be denoted by $(r,s) \in \{1,..,N_1\}\times\{1,..,N_2\}$.

The stock increase and decrease moments and the expected lengths of stock increase and decrease periods can all be calculated as "costs" till absorption in a certain subset of states, starting from another set.

For instance, the expected stock increase EU^i can be calculated as the sum of all contributions (increases) till absorption in a "decreasing" machine state (r,s), that is a state with $\nu_r < \varphi_s$, starting from some "increasing" state (i,j), that is with $\nu_i > \varphi_j$. To get the total expected stock increase, we have to weigh the states (i,j) with the conditional steady state probability $\pi_{(i,j)}$ that the increase starts in machine state (i,j). The sets needed consist of "increase", "decrease" and "neutral" states, and are denoted by D_+, D_- and D_0, respectively. More precisely, the sets D_+, D_- and D_0 are defined by

$$D_+ = \{(i,j) \in \{1,..,N_1\}\times\{1,..,N_2\} \mid \nu_i > \varphi_j\}$$
$$D_0 = \{(i,j) \in \{1,..,N_1\}\times\{1,..,N_2\} \mid \nu_i = \varphi_j\}$$
$$D_- = \{(i,j) \in \{1,..,N_1\}\times\{1,..,N_2\} \mid \nu_i < \varphi_j\}$$

The cardinalities of these sets are d_+, d_0 and d_-, respectively. The states of the machine system are ordered in the sequence D_+, D_0, D_- and within these sets in increasing lexicographic order. Note that this ordering differs from the ordering in section 2.4. In Fig. 2.6.2 these sets are sketched with possible transitions from one set to another.

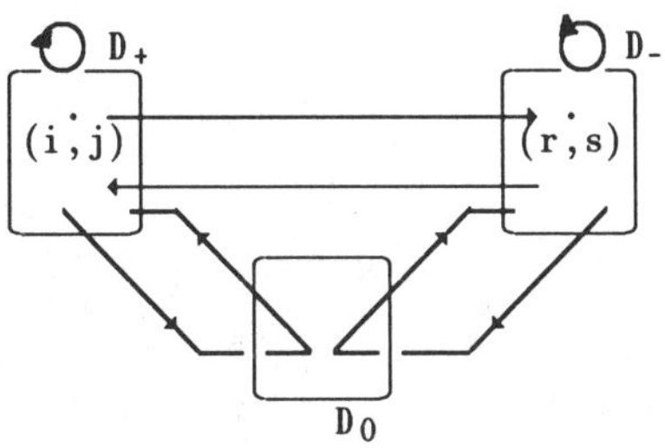

Figure 2.6.2.Transitions between states in D_+, D_0 and D_-.

Let

$f_{(i,j)}$ be the "cost" till absorption in a state of D_-, when we are now in state $(i,j) \in D_+ \cup D_0$.

$g_{(r,s)}$ be the cost till absorption in a state of D_+, when we are now in state $(r,s) \in D_0 \cup D_-$.

$\alpha_{(i,j)}$ be the cost rate in state $(i,j) \in D_+ \cup D_0 \cup D_-$

EC_{D_-} be the expected total cost till absorption in a state in D_-, when starting in D_+.

EC_{D_+} be the expected total cost till absorption in a state in D_+, when starting in D_-.

The expected cost till absorption in a state of D_- is found by weighing the contributions $Ef_{(i,j)}$ for each state (i,j) in D_+ by the conditional probability that an increase period starts in that state, $\pi_{(i,j)}$. Hence, EC_{D_-} can be written as

$$EC_{D_-} = \sum_{(i,j)\in D_+} Ef_{(i,j)}\,\pi_{(i,j)}, \tag{2.39}$$

where $\pi_{(i,j)}$, for $(i,j)\in D_+$, is the conditional probability that an increase period is started in state (i,j). $\sum_{(i,j)\in D_+} \pi_{(i,j)}=1$, $\pi_{(i,j)} \geq 0$.

In the same way

$$EC_{D_+} = \sum_{(r,s)\in D_-} Eg_{(r,s)}\,\pi_{(r,s)} \tag{2.40}$$

with $\sum_{(r,s)\in D_-} \pi_{(r,s)}=1$, $\pi_{(r,s)} \geq 0$.

In the same way $EC^2_{D_+}$ and $EC^2_{D_-}$ can be defined. The first and second moments of stock increase and stock decrease can then be determined from EC_{D_+}, EC_{D_-}, $EC^2_{D_+}$ and $EC^2_{D_-}$, respectively, by choosing the right values for the cost rates $\alpha_{(i,j)}$. EL^i and EL^d can also be determined from EC_{D_+} and EC_{D_-}, respectively, by appropriate choices for $\alpha_{(i,j)}$. These values of $\alpha_{(i,j)}$ are indicated in Table 2.6.1.

Note that in measuring stock increase and decrease, states in D_0 are not absorbing. Measuring stock increase is only stopped if a state in D_- is entered and measuring stock decrease is only stopped if a state in D_+ is entered.

$(i,j)\in$	D_+	D_0	D_-
EU^i	$\nu_i-\varphi_j$	0	–
$E(U^i)^2$	$\nu_i-\varphi_j$	0	–
EU^d	–	0	$\varphi_j-\nu_i$
$E(U^d)^2$	–	0	$\varphi_j-\nu_i$
EL^i	1	1	–
EL^d	–	1	1

Table 2.6.1. Values of $\alpha_{(i,j)}$ corresponding with some buffer quantities.

Hence, in order to determine EC_{D_+}, $EC^2_{D_+}$, EC_{D_-} and $EC^2_{D_-}$ and consequently EU^i, $E(U^i)^2$, EU^d, $E(U^d)^2$, EL^i and EL^d, we have to proceed in two steps.

1. Determination of the costs $Ef_{(i,j)}$, $Ef^2_{(i,j)}$, $Eg_{(r,s)}$, $Eg^2_{(r,s)}$.
2. Determination of the weight vector π (of dimension $d_+ + d_-$).

1. Determination of $Ef_{(i,j)}$, $Ef^2_{(i,j)}$, $Eg_{(r,s)}$, $Eg^2_{(r,s)}$

The cost till absorption in a state of D_-, starting from an increasing or neutral state (i,j), $f_{(i,j)}$, can be calculated by splitting up the cost in the cost in state (i,j) and the cost after a transition has been made to another increasing or neutral state. For $(i,j) \in D_+ \cup D_0$ we obtain

$$Ef_{(i,j)} = \frac{\alpha_{(i,j)}}{\lambda_i+\mu_j} + \frac{\lambda_i}{\lambda_i+\mu_j} \sum_{\{u\,|\,(u,j)\in D_+ \cup D_0\}} p_{iu} Ef_{(u,j)} + \frac{\mu_j}{\lambda_i+\mu_j} \sum_{\{v\,|\,(i,v)\in D_+ \cup D_0\}} q_{jv} Ef_{(i,v)} \tag{2.41}$$

After $Ef_{(i,j)}$ has been calculated it is possible to obtain $Ef^2_{(i,j)}$ in a similar way.

$$Ef^2_{(i,j)} = \frac{2\alpha^2_{(i,j)}}{(\lambda_i+\mu_j)^2} + \frac{2\alpha_{(i,j)}}{\lambda_i+\mu_j} \left[\frac{\lambda_i}{\lambda_i+\mu_j} \sum_{\{u\,|\,(u,j)\in D_+ \cup D_0\}} p_{iu} Ef_{(u,j)} + \frac{\mu_j}{\lambda_i+\mu_j} \sum_{\{v\,|\,(i,v)\in D_+ \cup D_0\}} q_{jv} Ef_{(i,v)}\right] + \frac{\lambda_i}{\lambda_i+\mu_j} \sum_{\{u\,|\,(u,j)\in D_+ \cup D_0\}} p_{iu} Ef^2_{(u,j)} + \frac{\mu_j}{\lambda_i+\mu_j} \sum_{\{v\,|\,(i,v)\in D_+ \cup D_0\}} q_{jv}\, Ef^2_{(i,v)} \tag{2.42}$$

Similarly, we have for the $g_{(r,s)}$'s, for $(r,s) \in D_0 \cup D_-$

$$Eg_{(r,s)} = \frac{\alpha_{(r,s)}}{\lambda_r+\mu_s} + \frac{\lambda_r}{\lambda_r+\mu_s} \sum_{\{u|(u,s)\in D_0 \cup D_-\}} p_{ru} Eg_{(u,s)} + \frac{\mu_s}{\lambda_r+\mu_s} \sum_{\{v|(r,v)\in D_0 \cup D_-\}} q_{sv} Eg_{(r,s)} \tag{2.43}$$

$$Eg^2_{(r,s)} = \frac{2\alpha^2_{(r,s)}}{(\lambda_r+\mu_s)^2} + \frac{2\alpha_{(r,s)}}{\lambda_r+\mu_s} \left[\frac{\lambda_r}{\lambda_r+\mu_s} \sum_{\{u|(u,s)\in D_0 \cup D_-\}} p_{ru} Eg_{(u,s)} + \frac{\mu_s}{\lambda_r+\mu_s} \sum_{\{v|(r,v)\in D_0 \cup D_-\}} q_{sv} Eg_{(r,v)}\right] + \frac{\lambda_r}{\lambda_r+\mu_s} \sum_{\{u|(u,s)\in D_0 \cup D_-\}} p_{ru} Eg^2_{(u,s)} + \frac{\mu_s}{\lambda_r+\mu_s} \sum_{\{v|(r,v)\in D_0 \cup D_-\}} q_{sv} Eg^2_{(r,v)} \tag{2.44}$$

Let the $d_+ + d_0$ vector Ef be given by $(Ef_{(i,j)})^T_{(i,j)\in D_+ \cup D_0}$. Ef^2, Eg and Eg^2 are defined similarly. The system (2.41) can be written as

$$Ef = \mathbf{a} + \mathbf{A_1} Ef, \tag{2.45}$$

where **a** is a $d_+ + d_0$ vector. Similarly,

$$Ef^2 = \mathbf{a^*} + \mathbf{A_1} Ef^2, \tag{2.46}$$

The matrix $(\mathbf{I} - \mathbf{A_1})$ is non-singular. This follows, by Theorem 2.1, corollary 3 in Seneta (1973), from the fact that $\mathbf{A_1}$ is a an irreducible, substochastic matrix. $\mathbf{A_1}$ being substochastic means that $A_1(i,j) \geq 0$, $\sum_j A_1(i,j) \leq 1$ for all i and $\sum_j A_1(i,j) < 1$, for some i. The irreducibility of $\mathbf{A_1}$ follows from the irreducibility of **P** and **Q**. Hence $Ef = (\mathbf{I} - \mathbf{A_1})^{-1}\mathbf{a}$, $Ef^2 = (\mathbf{I} - \mathbf{A_1})^{-1}\mathbf{a^*}$. The vector Ef^2 can be calculated after Ef has been determined. Similarly $\mathbf{Eg} = (\mathbf{I} - \mathbf{A_2})^{-1}\mathbf{b}$, $Eg^2 = (\mathbf{I} - \mathbf{A_2})^{-1}\mathbf{b^*}$, where the $(d_0+d_-)\times(d_0+d_-)$ matrix $\mathbf{A_2}$ and the d_0+d_- vectors **b** and $\mathbf{b^*}$ can be determined from (2.43) and (2.44).

2. Determination of π.

To determine the steady state probabilities $\pi_{(i,j)}$ we proceed as follows.
Let the $(d_+ + d_0)\times d_-$ matrix $\mathbf{\Theta}^*$ and the $(d_0+d_-)\times d_+$ matrix $\mathbf{\Theta}$ be defined as follows

$\theta^*_{((i,j),(r,s))}$ = conditional probability of absorption in decreasing state $(r,s)\in D_-$, given that the start is in a nondecreasing state $(i,j) \in D_+ \cup D_0$.

$\theta_{((r,s),(i,j))}$ = conditional probability of absorption in increasing state $(i,j)\in D_+$, given that the start is in nonincreasing state $(r,s) \in D_0 \cup D_-$.

We have the following relations

$$\theta^*_{((i,j),(r,s))} = \frac{\lambda_i}{\lambda_i+\mu_j} \sum_{\{u \mid (u,j) \in D_+ \cup D_0\}} p_{iu}\theta^*_{((u,j),(r,s))} + \frac{\mu_j}{\lambda_i+\mu_j} \sum_{\{v \mid (i,v) \in D_+ \cup D_0\}} q_{jv}\theta^*_{((i,v),(r,s))} , \quad \text{for } i \neq r,\ j \neq s$$

$$\theta^*_{((r,j),(r,s))} = \frac{\mu_j q_{js}}{\lambda_r+\mu_j} + \frac{\lambda_r}{\lambda_r+\mu_j} \sum_{\{u \mid (u,j) \in D_+ \cup D_0\}} p_{ru}\theta^*_{((u,j),(r,s))} + \frac{\mu_j}{\lambda_r+\mu_j} \sum_{\{v \neq s \mid (r,v) \in D_+ \cup D_0\}} q_{jv}\theta^*_{((r,v),(r,s))} , \quad \text{for } j \neq s$$

$$\theta^*_{((i,s),(r,s))} = \frac{\lambda_i p_{ir}}{\lambda_i+\mu_s} + \frac{\lambda_i}{\lambda_i+\mu_s} \sum_{\{u \neq r \mid (u,s) \in D_+ \cup D_0\}} p_{iu}\theta^*_{((u,s),(r,s))} + \frac{\mu_s}{\lambda_i+\mu_s} \sum_{\{v \mid (i,v) \in D_+ \cup D_0\}} q_{sv}\theta^*_{((i,v),(r,s))} , \quad \text{for } i \neq r$$

The $\theta_{((r,s),(i,j))}$'s satisfy similar equations

$$\theta_{((r,s),(i,j))} = \frac{\lambda_r}{\lambda_r+\mu_s} \sum_{\{u \mid (u,s) \in D_0 \cup D_-\}} p_{ru}\theta_{((u,s),(i,j))} + \frac{\mu_s}{\lambda_r+\mu_s} \sum_{\{v \mid (r,v) \in D_0 \cup D_-\}} q_{sv}\theta_{((r,v),(i,j))} , \quad \text{for } r \neq i,\ s \neq j$$

$$\theta_{((i,s),(i,j))} = \frac{\mu_s q_{sj}}{\lambda_i+\mu_s} + \frac{\lambda_i}{\lambda_i+\mu_s} \sum_{\{u \mid (u,s) \in D_0 \cup D_-\}} p_{iu}\theta_{((u,s),(i,j))} + \frac{\mu_s}{\lambda_i+\mu_s} \sum_{\{v \neq j \mid (i,v) \in D_0 \cup D_-\}} q_{sv}\theta_{((i,v),(i,j))} , \quad \text{for } s \neq j$$

$$\theta_{((r,j),(i,j))} = \frac{\lambda_r p_{ri}}{\lambda_r+\mu_j} + \frac{\lambda_r}{\lambda_r+\mu_j} \sum_{\{u \neq i \mid (u,j) \in D_0 \cup D_-\}} p_{ru}\theta_{((u,j),(i,j))} + \frac{\mu_j}{\lambda_r^1+\lambda_j^2} \sum_{\{v \mid (r,v) \in D_0 \cup D_-\}} q_{jv}\theta_{((r,v),(i,j))} , \quad \text{for } r \neq i$$

Rewritten in matrix notation

$$\boldsymbol{\Theta}^* = \boldsymbol{\Theta}_1^* + A_1\boldsymbol{\Theta}^* \tag{2.47}$$

$$\boldsymbol{\Theta} = \boldsymbol{\Theta}_1 + A_2\boldsymbol{\Theta} , \tag{2.48}$$

where $\boldsymbol{\Theta}_1^*$ is a matrix of dimension $(d_+ + d_0) \times d_-$, $\boldsymbol{\Theta}_1$ is of dimension $(d_0 + d_-) \times d_+$ and $\mathbf{A_1}$ and $\mathbf{A_2}$ are the same matrices as before. The matrices $\boldsymbol{\Theta}$ and $\boldsymbol{\Theta}^*$ can be solved from (2.47) and (2.48).

$$\boldsymbol{\Theta}^* = (I - A_1)^{-1}\boldsymbol{\Theta}_1^*, \quad \boldsymbol{\Theta} = (I - A_2)^{-1}\boldsymbol{\Theta}_1$$

The weights $\pi_{(i,j)}$ needed for the calculation of the costs C_{D_-} and C_{D_+} can now be obtained from $\boldsymbol{\Theta}$ and $\boldsymbol{\Theta}^*$ as follows. Let $\tilde{\boldsymbol{\Theta}}^*$ be the $(d_+ \times d_-)$–matrix consisting of the first d_+ rows of $\boldsymbol{\Theta}^*$ and $\tilde{\boldsymbol{\Theta}}$ the $(d_- \times d_+)$–matrix consisting of the last d_- rows of $\boldsymbol{\Theta}$. The following relations exist for the weights $\pi_{(i,j)}$.
For $(i,j) \in D_+$

$$\pi_{(i,j)} = \sum_{(r,s) \in D_-} \pi_{(r,s)} \tilde{\theta}_{((r,s),(i,j))},$$

This means that the probability of starting an increase period in state (i,j) equals the sum of the probabilities of starting in decrease state (r,s), times the probability of absorption in state (i,j), given that we start in state (r,s). Note that a *stock increase period can only start in D_+* and a *stock decrease period can only start in D_-*. Therefore we do not have to incorporate states in D_0.
Similarly we have for $(r,s) \in D_-$

$$\pi_{(r,s)} = \sum_{(i,j) \in D_+} \pi_{(i,j)} \tilde{\theta}^*_{((i,j),(r,s))},$$

which can be rewritten in matrix notation as

$$(\boldsymbol{\pi}^+)^T = (\boldsymbol{\pi}^+)^T \tilde{\boldsymbol{\Theta}}^* \tilde{\boldsymbol{\Theta}} \tag{2.49}$$

and

$$(\boldsymbol{\pi}^-)^T = (\boldsymbol{\pi}^-)^T \tilde{\boldsymbol{\Theta}} \tilde{\boldsymbol{\Theta}}^* , \tag{2.50}$$

with $\boldsymbol{\pi}^+ = (\pi_{(i,j)})_{(i,j) \in D_+}$, $\boldsymbol{\pi}^- = (\pi_{(r,s)})_{(r,s) \in D_-}$.
Under the normalizing conditions $1^T_{d_+} \cdot \boldsymbol{\pi}^+ = 1$ and $1^T_{d_-} \cdot \boldsymbol{\pi}^- = 1$ there exist unique solutions $\boldsymbol{\pi}^+$ and $\boldsymbol{\pi}^-$ to (2.49) and (2.50).

The first and second moments of C_{D_-} and C_{D_+} are now calculated as

$$EC_{D_-} = \sum_{(i,j) \in D_+} \pi_{(i,j)} Ef_{(i,j)} \tag{2.51}$$

$$EC_{D_+} = \sum_{(r,s) \in D_-} \pi_{(r,s)} Eg_{(r,s)} \tag{2.52}$$

$$EC^2_{D_-} = \sum_{(i,j)\in D_+} \pi_{(i,j)} Ef^2_{(i,j)} \tag{2.53}$$

$$EC^2_{D_+} = \sum_{(r,s)\in D_-} \pi_{(r,s)} Eg^2_{(r,s)} \tag{2.54}$$

By making the substitutions for $\alpha_{(i,j)}$, indicated in Table 2.6.1 in equations (2.41), EC_{D_+}, obtained from (2.51) equals either EU^i or EL^i. $EC^2_{D_+}$ can be obtained from (2.53) by making the substitutions for $\alpha_{(i,j)}$ indicated in Table 2.6.1 in (2.42). In the same way, EU^d and EL^d can be obtained from (2.51) and $E(U^d)^2$ can be obtained from (2.54), by making the indicated substitutions for $\alpha_{(i,j)}$ in (2.43) and (2.44), respectively. Third moments of U^i and U^d can be calculated in the same way.

There exist relationships between the mean drift of the system and the line throughput without buffer on the one hand and the first moments of increase and decrease of the buffer contents on the other hand. Such relationships are of importance, in the approximation of complex two–stage lines by simple lines. In section 2.7, they are used for obtaining the parameters in the approximating line. These relationships will be derived in subsection 2.6.3.

2.6.3 Relations between buffer stock moments and line characteristics

In this subsection some relationships will be demonstrated which are valuable in the approximation of two–stage lines.
The following relationship exists between the mean drift of the system $\sigma = \text{eff}_1 - \text{eff}_2$ and EU^i, EU^d, EL^i, EL^d.

$$\sigma = \frac{EU^i - EU^d}{EL^i + EL^d}. \tag{2.55}$$

Relation (2.55) is intuitively clear: the difference between the average machine speeds is equal to the difference between average stock increase and stock decrease, per increase/decrease step, divided by the average length of such an increase/ decrease step. It can be demonstrated formally as follows. The left hand side equals $\sum_{(i,j)\in D_+ \cup D_0 \cup D_-} (\nu_i - \varphi_j)\, w_{(i,j)}$, where the N_1N_2–vector $\mathbf{w}$ is the equilibrium probability for the machine states of the process (see equation (2.19)). The right hand side can be interpreted as "cost" per cycle divided by mean cycle length. The cost rate factor in state (i,j) is $\nu_i - \varphi_j$. The cycle is defined on the sets D_+, D_- and D_0. In order to see this, we define two additional states a and b. The state a represents all states in D_+, b represents the states in D_-. See Fig. 2.6.3.

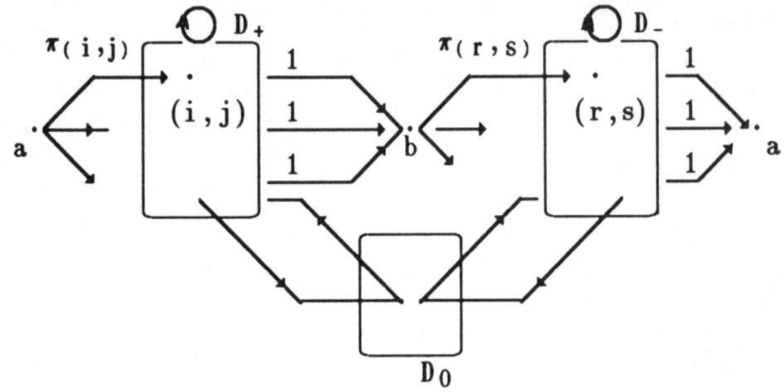

Figure 2.6.3. Transitions between states in D_+, D_0 and D_- with two additional states a and b.

Let T_a be the cost till absorption in state a, when starting in a, then

$$
\begin{aligned}
ET_a &= \sum_{(i,j)\in D_+} \pi_{(i,j)} \left[Ef_{(i,j)} + \sum_{(r,s)\in D_-} \pi_{(r,s)} Eg_{(r,s)} \right] \\
&= \sum_{(i,j)\in D_+} \pi_{(i,j)} Ef_{(i,j)} + \sum_{(r,s)\in D_-} \pi_{(r,s)} Eg_{(r,s)} \qquad (2.56)
\end{aligned}
$$

To obtain the mean cycle length from (2.56) we have to substitute 1 for the cost rate in state (i,j), $\alpha_{(i,j)}$, for all (i,j) $\in D_+ \cup D_0 \cup D_-$, in equations (2.41),(2.43). Using (2.51) and (2.52) it follows then that

$$ET_{a,1} = \text{mean cycle length} = EL^i + EL^d.$$

Substituting $\alpha_{(i,j)} = \nu_i - \varphi_j$, for all (i,j) $\in D_+ \cup D_0 \cup D_-$, in (2.41),(2.43), it follows, by using (2.51) and (2.52) once more

$$ET_{a,2} = \text{mean buffer drift per cycle} = EU^i - EU^d.$$

And hence by a renewal theoretical argument, it follows that the mean buffer drift equals $ET_{a,2}/ET_{a,1}$, which proves (2.55).

Similarly it can be shown that the throughput of the system with buffer of capacity 0, v(0), is related to EU^i, EU^d, EL^i and EL^d as follows

$$v(0) = eff_1 - \frac{EU^i}{EL^i+EL^d} = eff_2 - \frac{EU^d}{EL^i+EL^d} \tag{2.57}$$

To demonstrate the first part of (2.57), note that

$$v(0) = \sum_{(i,j)\in D_+ \cup D_0 \cup D_-} \min\{\nu_i,\varphi_j\} w_{(i,j)}, \tag{2.58}$$

where **w** is the equilibrium vector on the machine states (see (2.19)). Further eff_1 can be written as $\sum_{(i,j)\in D_+ \cup D_0 \cup D_-} \nu_i w_{(i,j)}$ and it can be shown that $\frac{EU^i}{EL^i+EL^d} = \sum_{(i,j)\in D_+} (\nu_i-\varphi_j) w_{(i,j)}$, by a similar argument as used in the demonstration of (2.55). The demonstration of the second part of (2.57) is similar to the first part.

Note that relation (2.58), and hence also relation (2.57), only holds for time-dependent rates. In case of state-dependent rates and no buffer, the rates of the exponential distributions in some states may change. Hence relation (2.57) will only be approximately true. The quality of the approximation depends on the type of state dependence. State-dependent rates are not further investigated here.

2.7 APPROXIMATION OF COMPLEX TWO-STAGE LINES

2.7.1 Introduction

In this section it is shown how complex two-stage lines, of the type described in section 2.4 can be approximated by simpler two-stage lines with the same first two (or three) moments of buffer stock increase and decrease. The reason for doing his is that the lines in section 2.4 may be difficult to analyze if the state space is large. Calculation of the first few moments of stock increase and decrease is computationally much simpler than calculation of the throughput, since the state space is reduced, by not incorporating the buffer state. The approximating line is chosen so, that calculation of the throughput, mean buffer contents and the moments of buffer stock increase and stock decrease is relatively simple. Results in this section are based on De Koster (1987a). The line of section 2.4 is sketched in

Fig. 2.7.1. It is assumed that all exponential rates are time–dependent.

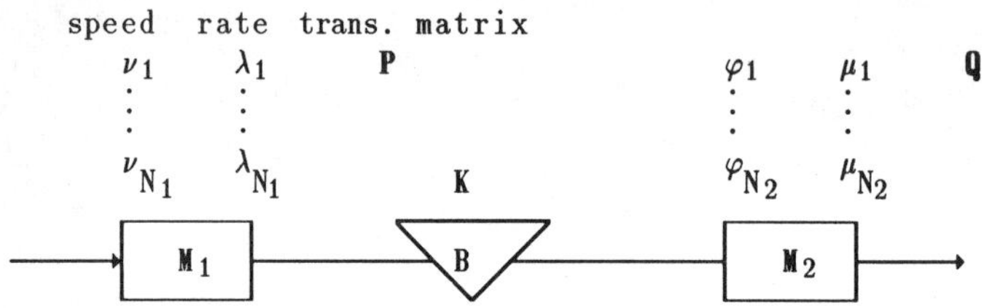

Figure 2.7.1. Two–stage line with many machine states

Instead of first and second moments of stock increase, it is equivalent to consider expectation and the coefficient of variation of buffer stock increase. The *coefficient of variation* (cv) of U^i is defined as

$$c^i = \frac{\sqrt{E(U^i)^2 - (EU^i)^2}}{EU^i} ,$$

The cv of U^d, c^d, is defined similarly.

The quantities EU^i, c^i, EU^d and c^d (and also higher moments of U^i and U^d, if necessary) of the line in Fig. 2.7.1 can be obtained as indicated in section 2.6. The aim is to find an as simple as possible line with the same first few moments of stock increase and stock decrease (also denoted by "stock moments"), and possibly with the same expected lengths of stock increase and stock decrease periods. In most cases, four states for the upstream machine and one state for the downstream machine will be shown to be sufficient, to obtain the same stock moments. In Fig. 2.7.2 a line is sketched, which will be used in this and the next subsection to approximate the line in Fig. 2.7.1.

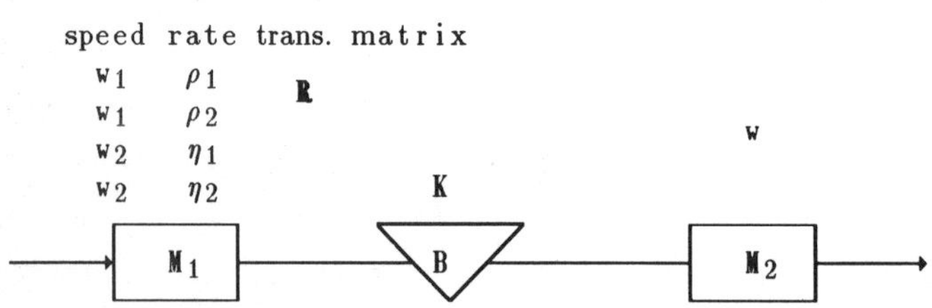

Figure 2.7.2. Two–stage line approximating the line in Fig. 2.7.1

The approximating line in Fig. 2.7.2 is a special case of the original line in Fig. 2.7.1, with $N_1 = 4$, $N_2 = 1$, $\lambda_i = \rho_i$, $\lambda_{i+2} = \eta_i$ (i=1,2), $\mu_1 = 0$, $\nu_1 = \nu_2 = w_1$, $\nu_3 = \nu_4 = w_2$ and $\varphi_1 = w$. Machine M_2 of the approximating line is completely reliable with production rate w. Machine M_1 only has two different speeds, $w_1 > w$ and $w_2 < w$, so that the buffer content sometimes increases and sometimes decreases. The transition matrix of M_1 is denoted by **R**. Note that the distribution of the length of a speed-w_1 or a speed-w_2 interval is the same as that of of the length of a buffer increase period or the length of a buffer decrease period, respectively. Furthermore, $EU^i_{ap}=(w_1-w)EL^i_{ap}$ and $EU^d_{ap}=(w-w_2)EL^d_{ap}$, where the index ap refers to the approximating line. This implies that the cv of the distribution of the length of a speed-w_1 interval equals c^i_{ap} and the cv (coefficient of variation) of the distribution of the length of a speed-w_2 interval equals c^d_{ap}. This fact can be used in using existing methods for fitting distributions on known first, second and third moments.

The problem is now to determine the parameters of the approximating line, so that its first moments of buffer increase and buffer decrease are the same as those of the line in Fig. 2.7.1. This will be demonstrated by means of an example line. The parameters of this example line are given in Table 2.7.1.

$N_1=5$ $N_2=1$	prod. rate	life rate	trans.matrix	EL^i+EL^d	EU^i	c^i	EU^d	c^d
origin. line	$\nu_1=1.5$	$\lambda_1=0.2$	$P = \begin{bmatrix} 0 & 0 & 0 & 0.2 & 0.8 \\ 0 & 0 & 0 & 0.9 & 0.1 \\ 0 & 0 & 0 & 0.9 & 0.1 \\ 0 & 0.1 & 0.9 & 0 & 0 \\ 0.2 & 0.3 & 0.5 & 0 & 0 \end{bmatrix}$	180.3	4.39	1.11	14.6	1.58
	$\nu_2=1.1$	$\lambda_2=0.2$						
	$\nu_3=1.05$	$\lambda_3=0.01$						
	$\nu_4=0.9$	$\lambda_4=0.01$						
	$\nu_5=0.01$	$\lambda_5=0.02$						
	$\varphi_1=1$	$\mu_1=0$						

Table 2.7.1 Parameters of the example line

Machine M_1 in this line has 5 states, machine M_2 has one state. The parameters $w_1, w_2, \rho_1, \rho_2, \eta_1, \eta_2$, the transition matrix **R** and the speed w of the approximating line in Fig 2.7.2 will be determined under the conditions:

$$EU^i_{ap}=EU^i,\ E(U^i_{ap})^2=E(U^i)^2, \tag{2.59a}$$

$$EU^d_{ap}=EU^d \text{ and } E(U^d_{ap})^2=E(U^d)^2. \tag{2.59b}$$

It is not strictly necessary to require that also

$$EL^i_{ap}=EL^i,\ EL^d_{ap}=EL^d. \tag{2.60}$$

By relation (2.38), it follows that these expectations can be incorporated later, to scale the approximate throughput correctly. Nevertheless, it will be clear that some freedom is left in the parameter choice, since there are only four conditions and 6 parameters plus the matrix **R** to choose. Hence relation (2.60) can be satisfied, if desired so.

Since $c^i>1$ and $c^d>1$ in Table 2.7.1, the distributions of the length of a w_1–speed and a w_2–speed interval must have a cv which is greater than 1 (and equal to c^i and c^d, respectively). Since $c^i>1$, the w_1–speed interval is chosen to be hyperexponentially distributed, with phases as sketched in Fig. 2.7.3.

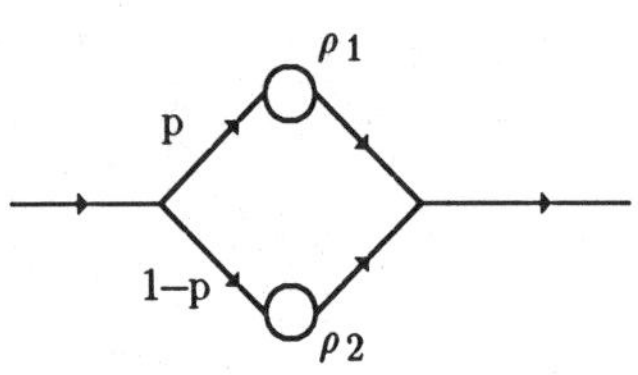

Figure 2.7.3. Two–phase hyperexponential distribution. Each phase has an exponential distribution (rates ρ_1 and ρ_2). p is the branching probability.

In case of a hyperexponentially distributed speed–w_1 interval with parameters as in Fig. 2.7.3, it follows that

$$E(U^i_{ap})^n = n!(w_1-w)^n\left(\frac{p}{\rho_1^n} + \frac{1-p}{\rho_2^n}\right),\ n \geq 1 \text{ and}$$

$$EL^i_{ap} = \frac{p}{\rho_1} + \frac{1-p}{\rho_2} \tag{2.61}$$

For w, $w_1 \geq w$ and p arbitrary, the rates of the hyperexponential w_1–interval determined by

$$\rho_1 = \frac{w_1-w}{EU^i}\left[1 + \left(\frac{1-p}{p}\,\frac{(c^i)^2-1}{2}\right)^{\frac{1}{2}}\right]^{-1}$$
$$\rho_2 = \frac{w_1-w}{EU^i}\left[1 - \left(\frac{p}{1-p}\,\frac{(c^i)^2-1}{2}\right)^{\frac{1}{2}}\right]^{-1} \tag{2.62}$$

make first and second moments of buffer increase equal (see Bolch and Akyildiz, 1982). This can be easily checked, by substituting the parameters obtained in (2.62) in formulas (2.61). The choice of p is arbitrary under the condition that $\rho_2>0$. Since for the choice of these five parameters there are only two conditions, namely those of (2.59a), other solutions are also possible. The choice of η_1 and η_2 is similar. If a speed–w_2 interval is hyperexponentially distributed with branching probability q, then the transition matrix **R** has the following form

$$\mathbf{R} = \begin{bmatrix} 0 & 0 & q & 1-q \\ 0 & 0 & q & 1-q \\ p & 1-p & 0 & 0 \\ p & 1-p & 0 & 0 \end{bmatrix}$$

Furthermore it is possible to make EL^i_{ap} and EL^i equal by choosing, for an arbitrary speed w, the speed w_1 such that

$$w_1-w = EU^i/EL^i \tag{2.63a}$$

EL^d_{ap} and EL^d can be made equal by choosing w_2 so that

$$w-w_2 = EU^d/EL^d \tag{2.63b}$$

The parameters w, w_1 and w_2 of the approximating line can be chosen so, that the throughputs of the lines are equal for the cases of infinite and zero buffer capacities. Assume that w is chosen so that $w=eff_{2,ap}:=eff_2$ and that the parameters w_1 and w_2 are chosen so that relations (2.63a) and (2.63b) hold. Assume furthermore that the other parameters of the approximating line are chosen so that conditions (2.59a) and (2.59b) are satisfied. From (2.63a) and (2.63b) it follows that condition (2.60) is satisfied. Together with the choice of w and relation (2.57), this implies

$$\hat{v}_{ap}(0) = v_{ap}(0) = v(0) \tag{2.64}$$

It furthermore follows from (2.59a), (2.59b), (2.60), (2.57) and the choice of w, that $eff_{1,ap} = eff_1$. Using that for an arbitrary two–stage line

$$v(\infty) = \min\{\mathrm{eff}_1,\mathrm{eff}_2\}$$

it follows that

$$\hat{v}_{ap}(\infty) = v_{ap}(\infty) = v(\infty) \tag{2.65}$$

In Table 2.7.2 the parameters of two different lines, "line a" and "line b", approximating the line in Table 2.7.1, are given. In both lines, the speed w is chosen equal to 1. In line a, condition (2.60) is not satisfied. In line b, condition (2.60) *is* satisfied, by choosing w_1 and w_2 according to (2.63a) and (2.63b), respectively. Hence $v_b(0)=v(0)=0.9190$ and $v_b(\infty)=v(\infty)=0.9434$, but $v_a(0)=0.8462$ and $v_a(\infty)=0.89245$.

	speeds	rates	branch. prob.	$EL^i_{ap}+EL^d_{ap}$
approx. line a	$w_1=1.2$	$\rho_1=0.034$ $\rho_2=0.069$	p=0.5	94.9
	$w_2=0.8$	$\eta_1=0.007$ $\eta_2=0.104$	q=0.5	
approx. line b	$w_1=1.05$	$\rho_1=0.086$ $\rho_2=0.0177$	p=0.5	180.267
	$w_2=0.85$	$\eta_1=0.0057$ $\eta_2=0.080$	q=0.5	

Table 2.7.2. Parameters of the approximating lines a and b

In Table 2.7.3 results of the approximation are given. The throughputs of the lines can be obtained by using results of section 2.4. For line a, the throughputs v(K) and $v_{ap,a}(K)$ cannot be compared directly, since condition (2.60) is not satisfied. Compared are v(K) and $\hat{v}_{ap,a}(K)$ and $\hat{v}_{ap,b}(K)$, where $\hat{v}_{ap,ab}(K)$ is defined as

$$\hat{v}_{ap,ab}(K) := v(\infty) - (v_{ap,ab}(\infty)-v_{ap,ab}(K))(EL^i_{ap,ab}+EL^d_{ap,ab})/(EL^i+EL^d) \tag{2.66}$$

This formula is based on (2.38). Note that for line b, $\hat{v}_{ap,b}(K)=v_{ap,b}(K)$. This follows from (2.64).

Relative errors in Table 2.7.3 are measured in the following way

$$\epsilon(K) = \frac{v(K) - \hat{v}_{ap}(K)}{v(\infty) - v(K)} \times 100\% \tag{2.67}$$

The reason for using this (strong) definition of relative error is that we want to measure the relative error in the loss of production due to a finite buffer rather than in the production rate itself.

K	v(K)	$\hat{v}_{ap,a}(K)$	$\hat{v}_{ap,b}(K)$	ϵ(%)	EX	$EX_{ap,a}$	$EX_{ap,b}$
0	0.9190	0.9190	0.9190	0	0	0	0
5	0.9333	0.9320	0.9320	12.9	1.904	0.417	1.920
10	0.9385	0.9368	0.9368	34.7	3.181	0.624	3.281
20	0.9420	0.9409	0.9409	78.6	4.625	0.737	4.956
30	0.9430	0.9424	0.9424	150.0	5.238	0.751	5.819
40	0.9433	0.9430	0.9430	300.0	5.476	0.752	6.254
∞	0.9434	0.9434	0.9434	–	–	–	–

Table 2.7.3. Relative errors in approximation of the line in Table 2.7.1, by the lines in Table 2.7.2

From results in Table 2.7.3 it appears that $\hat{v}_{ap,a}(K)=\hat{v}_{ap,b}(K)$, although the lines are different and $EL_a^i+EL_a^d \neq EL_b^i+EL_b^d$. This result holds more general.
Let "line a" and "line b" be two different approximating lines with hyper-exponential w_1-interval and w_2-interval. From formulas (2.62) and its analogue with η, q and w–w_2 instead of ρ, p and w_1–w, it follows that if

$$\frac{\rho_{ia}}{w_{1a}-w} = \frac{\rho_{ib}}{w_{1b}-w} \quad i = 1,2$$
$$\frac{\eta_{ia}}{w-w_{2a}} = \frac{\eta_{ib}}{w-w_{2b}} \quad i = 1,2$$
$$p_a = p_b$$
$$q_a = q_b \tag{2.68}$$

we have $E(U_a^i)^n = E(U_b^i)^n$ and $E(U_a^d)^n = E(U_b^d)^n$ for all n. Relations (2.68) are also satisfied by the parameters of the two lines in Table 2.7.2. In section 2.6, it was argued that if all buffer increase and buffer decrease moments are the same in the two lines, then the scaled throughputs are equal. In formula:

$$(v_{ap,a}(\infty) - v_{ap,a}(K))(EL_a^i + EL_a^d) = (v_{ap,b}(\infty) - v_{ap,b}(K))(EL_b^i + EL_b^d) \tag{2.69}$$

In De Koster and Wijngaard (1985) relation (2.69) is proven for the case where $\rho_1 = \rho_2$, $\eta_1 = \eta_2$, w = 1 and w_2 = 0. Relation (2.69) implies that the choice of w, w_1 and w_2, p and q has no effect on the quality of the *throughput* approximation, as long as conditions (2.68) are satisfied. Therefore, for the throughput approximation we do not necessarily have to use the existing freedom to make EL^i+EL^d and $EL^i_{ap}+EL^d_{ap}$ equal.

In Table 2.7.3 also the mean buffer contents (EX) are given. They can be obtained using results of sections 2.3 and 2.4. For the mean buffer contents approximation, it may be important to choose the mean lengths of increase and decrease periods equal, as can be seen from results in Table 2.7.3. There appears to be a difference between the situation where (2.60) is satisfied and the situation where this is not the case.

Relative errors in throughput in Table 2.7.3 are large, but this is not serious since $v(K) - \hat{v}_{ap}(K)$ is small anyway. These large errors are also caused by the fact that in the numerical solution procedure the fourth digit is not always correct. However, the differences in EX and $EX_{ap,b}$ are more serious. In view of the fairly large errors, also a three-moment approximation of the line was used. Third moments of buffer increase and buffer decrease can be obtained by using known first and second moments, just as first moments are used to calculate second moments (see section 2.6.2).

In Altiok (1985) a method is presented for fitting a distribution to the first three moments of a distribution with cv > 1, using a two-phase Coxian-distribution (see also Whitt, 1982). A two-phase Coxian distribution is sketched in Fig 2.7.4.

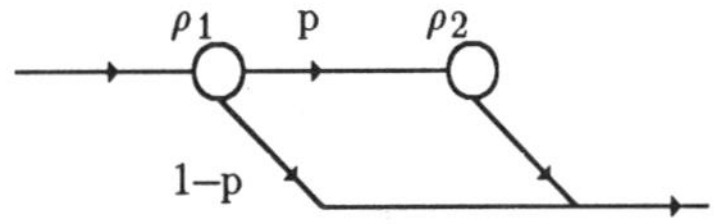

Figure 2.7.4. Two-phase Coxian distribution; branching probability p, rates ρ_1 and ρ_2.

If a speed-w_1 interval is two-phase Coxian distributed, it can be shown that

$$E(U^i_{ap})^n = n!(w_1-w)^n\left(\frac{1}{\rho_1^n} + p\sum_{k=1}^{n}\left(\frac{1}{\rho_2}\right)^k\left(\frac{1}{\rho_1}\right)^{n-k}\right) , \; n \geq 1$$

$$EL^i_{ap} = \frac{1}{\rho_1} + \frac{p}{\rho_2} \qquad (2.70)$$

A two-phase Coxian distribution may have a cv greater than 1 or smaller than 1. From the results in Altiok (1985) it follows that it is possible to keep the first three moments of buffer increase in the approximating line equal to the first three moments of buffer increase in the original line, by choosing, for w and $w_1 > w$ arbitrary,

$$\rho_1 = (X + \sqrt{X^2-4Y})/2$$
$$\rho_2 = X - \rho_1$$
$$p = \frac{\rho_2}{\rho_1}(m_1\rho_1-1),$$

where $Y=(6m_1-3m_2/m_1)/[3m_2^2/(2m_1)-m_3]$, $X=1/m_1+m_2Y/(2m_1)$ and $m_n=\frac{ET^n}{(w_1-w)^n}$, n=1,2,3.

This can be checked, by substituting these parameters in formulas (2.70). Note however, that a three-moment approximation by a two-phase Coxian distribution is only possible if $c^i>1$ and if

$$3m_2^2 < 2m_1m_3. \tag{2.71}$$

If $c^d>1$, η_1, η_2 and q are determined similarly for arbitrary $w_2 < w$. Instead of w_1-w we have to take $w-w_2$.

The parameters of the line in Fig. 2.7.2, approximating the line in Fig. 2.7.1 are now chosen by this three-moment method. Relative errors of this approximation are given in Table 2.7.4. The production rate w is again chosen to be equal to 1.0, EL^i and EL^d are chosen equal in the two lines.

K	v(K)	$\hat{v}_{ap}(K)$	ϵ(%)	EX	EX_{ap}	mach. speed	life rate	trans. prob.	EL+EM
0	0.9190	0.9190	0	0	0	w_1=1.051	ρ_1=0.084	p=0.757	180.267
5	0.9333	0.9333	−0.4	1.904	1.808		ρ_2=0.010		
10	0.9385	0.9386	−1.5	3.181	2.955	w_2=0.845	η_1=0.015	q=0.093	
20	0.9420	0.9421	−4.5	4.625	4.205		η_2=0.003		
30	0.9430	0.9430	−7.8	5.237	4.717				
40	0.9433	0.9433	−11.0	5.476	4.911				
∞	0.9434	0.9434	–	–	–				

Table 2.7.4. Approximations of the line in Table 2.7.1 by three-moment method.

The relative errors in the throughput in Table 2.7.4 are much smaller than the relative errors in Table 2.7.3. Also the differences in mean buffer content of the lines have become slightly smaller.

This example shows, how a given line as in Fig 2.7.1 can be approximated for the throughput and mean buffer content, by requiring that relations (2.59a)–(2.59b) are satisfied. If necessary, also third moments of buffer increase and buffer decrease can be included. The above method is generally applicable in case $c^i>1$ and $c^d>1$, if the condition (2.71) is satisfied. If this condition is not satisfied, it can be relaxed such that it is nearly satisfied. See Altiok (1985). If $c^i \leq 1$ or $c^d \leq 1$, we have to proceed differently. Here only the case $c^i \leq 1$ is treated. The case $c^d \leq 1$ is similar.

We distinguish two cases
1. $1/2 \leq (c^i)^2 < 1$
2. $(c^i)^2 < 1/2$

If $1/2 \leq (c^i)^2 < 1$, the structure of the approximating line can again be chosen as indicated in Fig. 2.7.2. A two–phase Coxian distribution as sketched in Fig. 2.7.4 can be chosen for the speed–w_1 interval distribution. In order to meet the conditions (2.59a), the parameters of the two–phase Coxian distribution may be determined as follows: (see Sauer and Chandy, 1981)

w and w_1 arbitrary

$$p = 1 - \frac{2(c^i)^2 - (2-2(c^i)^2)^{\frac{1}{2}}}{(c^i)^2 + 1}$$

$$\rho_1 = \rho_2 = \frac{w_1 - w}{EU^i}(1+p) \tag{2.72}$$

That the two conditions in (2.59a) are satisfied can be checked by substituting the choice of (2.72) in formulas (2.70).

For the case $(c^i)^2<1/2$, the structure of the distribution in Fig. 2.7.4 is no longer sufficient. More phases are needed for the speed–w_1 interval. The distribution of this interval can be taken of the kind sketched in Fig. 2.7.5.

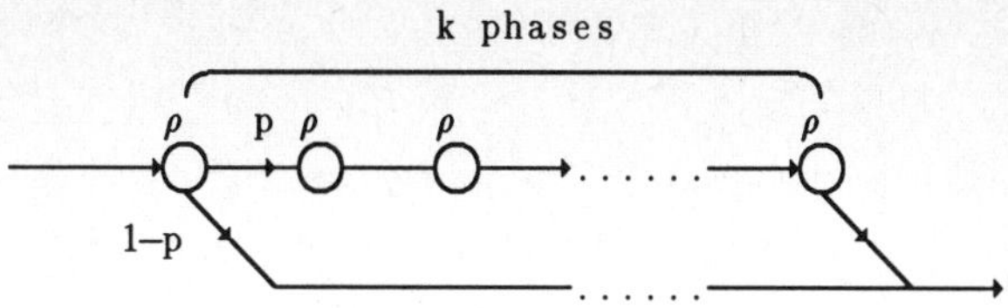

Figure 2.7.5. k–phase distribution

Note that a two–phase Coxian distribution with $\rho_1=\rho_2=\rho$ is the same as a two–phase distribution of the type sketched in Fig. 2.7.5. The number of phases in the distribution in Fig. 2.7.5, k, should be such that

$$\frac{1}{k} \leq (c^i)^2 \leq \frac{1}{k-1} \tag{2.73}$$

Once k is found from (2.73) the parameters p and ρ can be found from the formulas

$$\begin{aligned} p &= 1 - \frac{2k(c^i)^2 + k-2 - (k^2+4-4k(c^i)^2)^{\frac{1}{2}}}{2((c^i)^2+1)(k-1)} \\ \rho &= \frac{w_1-w}{EU^i}\,(p + k(1-p)) \end{aligned} \tag{2.74}$$

See also Marie (1980).

For all these phase–type distributions, the exact shape of the transition matrix **R** can be derived from the type of distribution used for the speed–w_1 and speed–w_2 interval. It does not depend on the rates in the phases. For example, if a speed–w_1 interval is hyperexponentially distributed with branching probability p and a w_2–interval is Coxian–distributed with branching probability q then the transition matrix **R** has the following form

$$\mathbf{R} = \begin{bmatrix} 0 & 0 & 1 & 0 \\ 0 & 0 & 1 & 0 \\ (1-q)p & (1-q)(1-p) & 0 & q \\ p & 1-p & 0 & 0 \end{bmatrix}$$

In section 2.7.2, some more examples are treated. In all cases $(c^i)^2$ and $(c^d)^2$ are greater than 1, so that two–phase hyperexponential or Coxian distributions suffice.

2.7.2 More numeric results

In this section three lines of the type sketched in Fig. 2.7.1 are approximated for the throughput by lines of the type sketched in Fig. 2.7.2, by using relationships (2.59a) and (2.59b). These cases demonstrate the quality of the approximation for lines with different complexity. In the three different cases, machine M_1 has 3, 5 and 7 states, respectively. Machine M_2 has two states in all cases. Hence, the total number of machine states increases from 6 to 14.

$N_2 = 2$	$N_1=3$, $m=1$	$N_1=5$, $m=2$	$N_1=7$, $m=3$
original line	$\mathbf{P} = \begin{bmatrix} 0 & 0 & 1 \\ 0 & 0 & 1 \\ 0.5 & 0.5 & 0 \end{bmatrix}$ $c^i = 1.455$ $c^d = 1.001$	$\mathbf{P} = \begin{bmatrix} 0 & 1 & 0 & 0 & 0 \\ 0 & 0 & 0 & 0 & 1 \\ 0 & 0 & 0 & 1 & 0 \\ 0 & 0 & 0 & 0 & 1 \\ 0.5 & 0 & 0.5 & 0 & 0 \end{bmatrix}$ $c^i = 1.206$ $c^d = 1.001$	$\mathbf{P} = \begin{bmatrix} 0 & 1 & 0 & 0 & 0 & 0 & 0 \\ 0 & 0 & 1 & 0 & 0 & 0 & 0 \\ 0 & 0 & 0 & 0 & 0 & 0 & 1 \\ 0 & 0 & 0 & 0 & 1 & 0 & 0 \\ 0 & 0 & 0 & 0 & 0 & 1 & 0 \\ 0 & 0 & 0 & 0 & 0 & 0 & 1 \\ 0.5 & 0 & 0 & 0.5 & 0 & 0 & 0 \end{bmatrix}$ $c^i = 1.110$ $c^d = 1.001$
	$\nu_r = 1.25$, $r = 1,..,m$; $\lambda_r = m \times 0.01$, $r = 1,..,N_1-1$ $\nu_j = 0.8$, $j = m+1,..,N_1-1$ $\nu_{N_1} = 0.01$; $\lambda_{N_1} = 0.09$		
approxim. line	a) $w_1=1.38$ $\rho_1=0.010$ $\rho_2=0.070$ $w_2=0.66$ $\eta_1=0.018$ $\eta_2=0.019$ $p=0.5$, $q=0.5$ $EL^i+EL^d=111.11$	$w_1=1.25$ $\rho_1=0.008$ $\rho_2=0.022$ $w_2=0.66$ $\eta_1=0.018$ $\eta_2=0.019$ $p=0.5$, $q=0.5$ $EL^i+EL^d=141.04$	$w_1=1.38$ $\rho_1=0.013$ $\rho_2=0.027$ $w_2=0.66$ $\eta_1=0.018$ $\eta_2=0.019$ $p=0.5$, $q=0.5$ $EL^i+EL^d = 111.11$
	b) $w_1=1.38$ $\rho_1=0.055$ $\rho_2=0.010$ $w_2=0.66$ $\eta_1=0.051$ $\eta_2=0.018$ $p=0.369$, $q=0.642$ EL +EM =111.11		

Table 2.7.5. Three lines with approximating lines. w=1.0 for all approximating lines.

For all cases we have $EL^i + EL^d = 111.11$, $EU^i=21.51$, $EU^d=18.90$ and M_1 has average machine speed 0.9235. For M_2 we have for all cases $N_2 = 2$, $\varphi_1 = 1.0$, $\varphi_2 = 0$, $\mu_1 = 0.01$, $\mu_2 = 0.09$ and $\mathbf{Q}=\begin{bmatrix} 0 & 1 \\ 1 & 0 \end{bmatrix}$. The average machine speed of M_2 is

0.9. Variation coefficients (and hence second moments) of stock increase and stock decrease differ for these cases, as indicated in Fig. 2.7.5. The cv's c^i and c^d decrease with the number of states. Since $c^i>1$ and $c^d>1$ in all cases, hyperexponential distributions are used for the lengths of the speed-w_1 and speed-w_2 intervals in the approximating line. In one case, a three-moment fit by a two-phase Coxian distribution is applied, as in the example in section 2.7.1. In all approximating lines w is chosen 1.0.

The parameters of the three lines are given in Table 2.7.5. In this table the approximating lines are also given. For case 1 we have two approximating lines, one based on a two-moment approximation ("line a") and one based on a three-moment approximation ("line b"). In line b (and the original line) $(EU^i)^3$=164,767 and $(EU^d)^3$=40,602. In line a $(EU^i)^3$=159,703 and $(EU^d)^3$=40,627.

Numerical results of the approximations are given in Table 2.7.6. $\hat{v}_{ap}(K)$ is defined by formula (2.66). Relative errors are measured with formula (2.67). We see that for case 1 approximation by line b, the first three moments keeping equal does not perform better than the approximation by line a, with only first and second moments kept equal.

	1					2			3		
K	v(K)	$\hat{v}_{ap,a}(K)$	ϵ	$\hat{v}_{ap,b}(K)$	ϵ	v(K)	$\hat{v}_{ap}(K)$	ϵ	v(K)	$\hat{v}_{ap}(K)$	ϵ
0	0.7299	0.7299	0	0.7299	0	.7299	.7299	0	.7299	.7299	0
5	0.7625	0.7626	−0.1	0.7633	−0.6	.7635	.7651	−1.1	.7637	.7655	−1.3
10	0.7828	0.7823	0.4	0.7837	−0.8	.7853	.7878	−2.2	.7860	.7890	−2.6
20	0.8078	0.8073	0.5	0.8097	−2.1	.8123	.8161	−4.3	.8140	.8182	−4.9
30	0.8234	0.8236	−0.3	0.8250	−2.1	.8291	.8332	−5.8	.8314	.8358	−6.4
40	0.8345	0.8353	−1.2	0.8365	−3.1	.8408	.8449	−6.9	.8434	.8476	−7.4
∞	0.9	0.9	–	0.9	–	.9	.9	–	.9	.9	–

Table 2.7.6. Relative errors (in %) in approximations of the lines in Table 2.7.5

From the results in this section it can be concluded that the throughput of two-stage lines with many machine states, like the ones sketched in Fig. 2.7.1, can be approximated very well by the throughput of simple lines, such as in Fig. 2.7.2, provided the first few moments of stock increase and stock decrease can be kept invariant. Two moments of buffer increase and buffer decrease are usually sufficient for an approximation of the throughput, but it is advised to include third moments if possible (see also Table 2.7.4.

In section 2.9, where the insertion department of section 2.1 is analyzed, the three–moment approximation method will be applied for obtaining estimates for throughput and mean buffer content.

2.8. DISCRETE PRODUCT MODELS

In the previous section it has been found that there is a high degree of insensitivity of the line performance for the exact machine parameters as long as expectation and variance of stock increase and stock decrease remain the same. In this section it will be demonstrated that there is also insensitivity with respect to continuity or discreteness of the goods flow. Some discrete–product two–stage lines with random service times will be compared with continuous product lines. A literature survey of such models was given in section 2.2. The approximation method can be used for discrete product lines by using stock increase and decrease moments. This possibility is evaluated in this section. The results in this section are not needed for the analysis of the insertion department case. They are based on De Koster and Wijngaard (1988b).

In the discrete product model considered here the assumptions are

1. M_1 has an exponentially distributed service time (parameter λ). M_2 has a two–phase Coxian distribution (see Fig. 2.7.4), with parameters μ_1, μ_2 and branching probability π.
2. M_1 does not start service if M_2 is servicing and the intermediate buffer is full (at level K).
3. M_1 has no storage capacity for products; that is products are transferred on service completion from the upstream to the downstream buffer. M_2 has capacity for 1 product at the operating position. It is equivalent to saying that M_2 has no capacity for products, but the buffer capacity is larger by one.

Note that this two–stage line is essentially an $M/C_2/1/K+1$ queueing system, where arrivals are lost if the content of the system is K+1.

In order to find a comparable continuous flow two–stage line, we have to set the upper bound of the intermediate buffer in such a continuous flow line equal to K+1 and the lower bound equal to 1. This is, because both the continuous flow system and the $M/C_2/1/K+1$ system then have a positive throughput for K=0. The continuous flow system is shifted–buffer equivalent (see section 2.5) with a system with buffer range [0,K]. The idea is now that, if we can construct a continuous flow line with approximately the same stock increase and buffer decrease, then the throughput and other performance measures will also be approximately the same. For such an approximation it is sufficient to keep two or three moments of stock increase and stock decrease and the frequency of such decrease and increase periods (for buffer contents approximations) equal.

For $M/C_2/1/K+1$ queueing systems, it is possible to calculate $E(U^i)^n$, $E(U^d)^n$ and EL^i, EL^d. This is done in appendix C, for n=1,2,3. The two–stage continuous flow system approximating the $M/C_2/1/K+1$ queueing system is the same as used in section 2.7. This line is sketched in Fig. 2.7.2.
By making EU^i, EU^d, EL^i and EL^d of the discrete and continuous product lines equal and by choosing $w = \frac{\mu_1\mu_2}{\pi\mu_1+\mu_2}$ (this is the production rate of M_2 in the discrete product system), it can easily be shown that

$$
\begin{aligned}
v_{dis}(0) &= v_{con}(0) \\
v_{dis}(\infty) &= v_{con}(\infty) \\
\lambda &= \text{eff}_1,
\end{aligned}
$$

where $v_{dis}(K)$ is the throughput of the $M/C_2/1/K+1$ queueing system, $v_{con}(K)$ is the throughput of the continuous flow line, and eff_1 is the net average machine speed of M_1. These equalities are analogues of (2.64), (2.65).

We will now give some numerical results for an M/M/1 and an $M/E_2/1$ queue (which are both special cases of the $M/C_2/1$ queue).

The increase and decrease of the M/M/1 queueing system can be obtained from the results listed in appendix C, by letting $\mu_1=\mu_2=:\mu$ and $\pi:=0$. The throughput and mean buffer content of the M/M/1/K+1 system can be found in every textbook on queueing systems (for example Kleinrock, 1975). Results for the throughput are given in Table 2.8.1. Relative differences (in %) are measured in the usual way, and not by formula (2.67).

K	$v_{dis}(K)$	$v_{con}(K)$	rel. difference (%)
0	0.474	0.474	0
1	0.631	0.641	−1.6
2	0.709	0.715	−0.8
5	0.808	0.810	−0.2
10	0.861	0.861	0
15	0.880	0.880	0
20	0.889	0.889	0
∞	0.9	0.9	

Table 2.8.1. Throughput of M/M/1/K+1 queueing system and approximating continuous flow line. The parameters are λ=0.9, μ=1, EU^i=0.9, EU^d=1.11, $E(U^i)^2$=2.52, $E(U^d)^2$=3.58, $E(U^i)^3$=10.13, $E(U^d)^3$=16.75, EL^i=1, EL^d=1.11. For the continuous flow line w_1=1.9, ρ_1=8.79, ρ_2=0.61, p=0.54, w_2=0, η_1=9.0, η_2=0.60, q=0.6, w=1.0 (notation as in Fig. 2.7.2).

As was expected, the relative differences are the greatest for low values of K, since there the discrete behavior of the system is dominant. Nevertheless all relative differences are slight.

It is more difficult to compare the mean buffer contents of both systems, since the total capacity of the queueing system for fixed K is one larger than that of the equivalent continuous flow system. If K=0, then $EX_{con}(K)$, the mean buffer content of the continuous flow system is 0, while $EX_{dis}(K)$ the mean *system* content of the discrete product system (this is the probability that M_2 is busy) is positive. Therefore, rather than comparing $EX_{dis}(K)$ and $EX_{con}(K)$ we compare $EX_{dis}(K)$–C and $EX_{con}(K)$, where the constant C is determined by $EX_{dis}(0)$–C = $EX_{con}(0)$ = 0. The constant C is the mean content of M_2 (or the utilization rate of M_2) in the case that there is no intermediate buffer. Results are listed in Table 2.8.2. Note that the relative differences in mean buffer content are also small.

K	$EX_{dis}(K)$–C	$EX_{con}(K)$	rel. diff.
0	0	0	0
1	0.456	0.459	−0.7
2	0.895	0.898	−0.4
5	2.109	2.111	−0.1
10	3.803	3.798	0.1
15	5.124	5.122	0.0
20	6.123	6.133	−0.2

Table 2.8.2. Comparison of mean buffer content for the lines with parameters as listed in Table 2.8.1. C=0.47

The second example is an $M/E_2/1/K+1$ queue. The moments of stock increase and stock decrease for this system can be obtained from results listed in appendix C by letting $\mu_1=\mu_2=:\mu$ and $\pi:=1$. The throughput and mean system content has been obtained by solving the balance equations, using the matrix geometric method (see Neuts, 1981). Results for the throughput and mean buffer content are listed in Table 2.8.3. The quantities $EX_{dis}(K)-C$ and $EX_{con}(K)$ are compared, instead of $EX_{dis}(K)$ and $EX_{con}(K)$. $C:=EX_{dis}(0)=0.48718$.

K	$v_{dis}(K)$	$v_{con}(K)$	rel.diff.	$EX_{dis}(K)-C$	$EX_{con}(K)$	rel.diff.
0	0.487	0.487	0	0	0	0
1	0.674	0.694	−3.0	0.477	0.476	0.2
2	0.760	0.770	−1.3	0.951	0.934	1.8
5	0.859	0.860	−0.1	2.298	2.251	2.0
10	0.908	0.907	0.1	4.306	4.233	1.7
20	0.935	0.938	−0.3	7.525	7.486	0.5
∞	0.95	0.95	0			

Table 2.8.3. Comparison of throughput of $M/E_2/1/K+1$ system and continuous flow system. $\lambda=0.95$, $\mu=2$, $EU^i=0.95$, $EU^d=1.05$, $E(U^i)^2=2.30$, $E(U^d)^2=2.84$, $E(U^i)^3=7.58$, $E(U^d)^3=10.99$, $EL^i=1.0$, $EL^d=1.05$, $w_1=1.95$, $w_2=0$, $w=1$, $\rho_1=2.59$, $\rho_2=0.69$, $q_1=0.42$, $\eta_1=4.84$, $\eta_2=0.70$, $q_2=0.59$. $C=0.48718$.

Note that the relative differences in throughput and mean buffer content in Table 2.8.3 are again slight, although they are greater than for the case of the M/M/1 queue.

In this section it has been shown that it is possible to approximate discrete product two–stage systems by continuous flow lines, with slight relative differences in throughput and mean buffer content. The mean system content of the discrete product system can not be compared straightforwardly with that of the continuous flow system. A correction has to be made for products within the server.

2.9 ANALYSIS OF THE INSERTION DEPARTMENT

In this chapter some tools have been developed for the analysis of continuous–product two–stage lines. In this section these tools are used to analyze the insertion department described in section 2.1. In particular, the results of sections

2.4, 2.6 and 2.7 will be exploited.

In section 2.1 it was noted, that batch sizes vary greatly and hence it is assumed that the length of a production run of an individual inserter, before switch–over or failure (i.e., the uptime) has an exponential distribution. Downtimes (due to repairs or switch–overs) are also assumed to be exponentially distributed. Each inserter can be in two states, up or down. The parameters of the insertion line which follow straightforwardly from Table 2.1.2 are given in Table 2.9.1. For instance, a repair rate .2767 for a horizontal inserter is equal to 1/3.61.

	machine speeds 10^3 hor.comp./ min.	failure rate λ	repair rate μ
horizontal ins.	2.875	0.03074	0.2767
vertical ins.	2.25	0.02289	0.1406

Table 2.9.1. Inserter parameters expressed in horizontal components

For the insertion department modeled in this way, the throughput as a function of K can be calculated numerically by using the analysis of section 2.4. The three horizontal inserters are equivalent with a single horizontal insertion machine having four different speeds (8.625, 5.75, 2.875 and 0). See also section 2.4.2. A speed 5.75–interval, for example, has an exponential distribution with a rate $2\lambda+\mu$, since only two inserters are operating. The transition matrix is also easy to calculate. The same can be done for the vertical inserters. Since the number of machine states is quite large (4×5) computational problems may arise. Therefore, the approximation by using stock increase/decrease moments, of section 2.7 is used. The original and resulting model are presented in Table 2.9.2. The notations in Table 2.9.2 are as in section 2.7. The buffer capacity in the model is denoted by K'. Recall that K' is obtained from the real workload norm K, by K'=K–9. Note that the mean system content in the real system is about 9 PCBs more than in the model. In the approximating line, the speed of the second machine (w) is 7.74, which equals the net production rate of vertical insertion (the net production rate of horizontal insertion is 7.77). The parameters of the first machine ($\rho_1,\rho_2,\eta_1,\eta_2$,p and q) are chosen by keeping the first three moments of buffer increase and buffer decrease equal in the approximating line and the original line. See section 2.7.1.

The parameters w_1 and w_2 are determined by requiring $EL^i_{ap}=EL^i$ and $EL^d_{ap}=EL^d$. As a result, $\hat{v}_{ap}(\infty)=v_{ap}(\infty)=v(\infty)$ and $\hat{v}_{ap}(0)=v_{ap}(0)= v(0)$. See formulas (2.64) and (2.65). Approximation results are given in Table 2.9.3.

$N_1=4$ $N_2=5$	prod. rate	life rate	trans. matrix	EL^i+EL^d	EU^i	c^i	EU^d	c^d
horiz. ins.	$\nu_1=8.63$ $\nu_2=5.75$ $\nu_3=2.88$ $\nu_4=0.0$	$\lambda_1=0.092$ $\lambda_2=0.33$ $\lambda_3=0.58$ $\lambda_4=0.83$	$P=\begin{bmatrix}0&1&0&0\\.82&0&.18&0\\0&.95&0&.05\\0&0&1&0\end{bmatrix}$	14.79	12.3	1.37	11.9	1.57
vertic. ins.	$\varphi_1=9.0$ $\varphi_2=6.75$ $\varphi_3=4.5$ $\varphi_4=2.25$ $\varphi_5=0.0$	$\mu_1=0.092$ $\mu_2=0.21$ $\mu_3=0.33$ $\mu_4=0.44$ $\mu_5=0.56$	$Q=\begin{bmatrix}0&1&0&0&0\\.67&0&.33&0&0\\0&.86&0&.14&0\\0&0&.95&0&.05\\0&0&0&1&0\end{bmatrix}$					
approx. line	$w_1=10.1$	$\rho_1=0.375$ $\rho_2=0.100$	$p=0.26$	14.79				
	$w_2=6.49$	$\eta_2=0.481$ $\eta_2=0.054$	$q=0.40$					
	$w=7.74$							

Table 2.9.2. Parameters of the insertion line and the approximating line

K'	$\hat{v}_{ap}(K')$	EX_{ap}
0	6.934	0
10	7.218	4.12
50	7.486	22.80
100	7.587	47.76
300	7.686	154.01
500	7.710	263.62
750	7.722	409.34
∞	7.737	

Table 2.9.3. Throughput (expressed in norm PCBs per minute) and mean buffer content (in norm PCBs) for the insertion line

These results can be used in the following way. Assume that the desired throughput is about 7.7 PCBs per minute. Then a way to realize this throughput is by keeping the WIP in the insertion department equal to about 500+9 PCBs, that is to about one hour of work for the insertion department as a whole. This workload corresponds to an average stock in the intermediate buffer of 264 PCBs.

The choice of exponential distributions for the life and repair times of the inserters may seem somewhat arbitrary. However, it will be clear that arbitrary distributions of phase type do not bring extra difficulties, in principle. If these distributions have many phases, some computational problems in the solution of the systems of linear equations for the determination of the moments of U^i and U^d may arise, so that combination of states is necessary. Also PCB–dependent speeds for the inserters can be handled. Then the modeling could be as indicated in section 2.4.2, example 3. These complications are not incorporated here. Product–dependent production rates will be discussed in Chapter 7.

2.10 CONCLUSIONS AND PREVIEW

In this chapter a model was introduced for the analysis and design of the capacity control level in two–stage production systems. Different aspects of the model have been highlighted, such as analytical tractability, influence of changes in model parameters and model assumptions. The model can also be used to approximate discrete product models. It appears to be simple but general enough for the modeling and analysis of two–stage production systems at capacity level.

In this chapter the emphasis has been on small, single–product production systems, with two stages. In the following chapters attention is shifted to more complex systems. Single–product systems with various layouts are studied in Chapters 3, 4 and 5. Multiproduct aspects are discussed in Chapters 6 and 7. For systems consisting of more than two stages, the control structure becomes important. Local control structures are studied in Chapters 3 and 4. In Chapters 5 and 6 integral control structures are studied.

CHAPTER 3

3 MORE COMPLEX SYSTEMS

In the previous chapter, a model has been introduced for the description of two–stage production systems at capacity level. Besides two–stage systems, there exist many other layouts for which the model must be used. This text concentrates on systems with layouts in which clear main product flows can be distinguished. Hence job shops are left out of scope. However, also in flow shop–like systems different layout elements can be distinguished, such as

- flow lines
- assembly and disassembly elements
- product type–dependent routings
- loops
- parallel branches and common buffers.

In Fig 3.1 some of these layout elements are sketched.

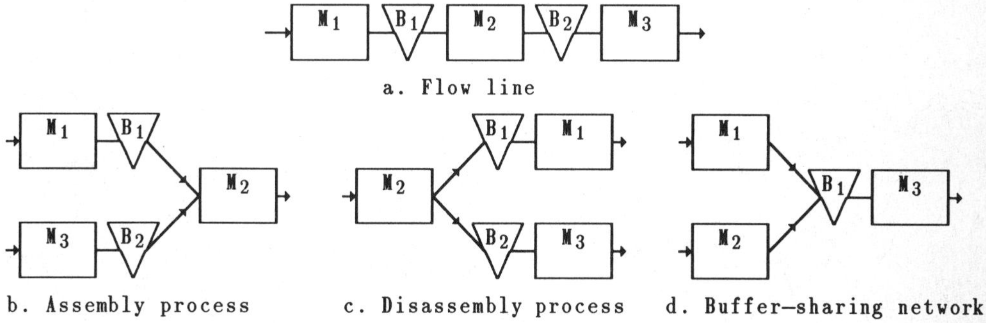

Figure 3.1. Examples of basic network layout elements

Different layouts studied in this monograph are discussed in section 3.1. In modeling production networks with layout elements as the ones mentioned above, it is assumed that only a single product type is produced. The product will later be interpreted as being an aggregate (Chapters 6 and 7). From the cases discussed in Chapter 7, it will become clear that the continuous–flow model can indeed be used for describing and analyzing production systems at capacity level. In section 3.1, available techniques for analyzing production networks are discussed. These techniques are analytic and numeric tools, heuristic methods and simulation. In the rest of this chapter, the emphasis is on analytic and numeric methods (heuristic techniques are considered in following chapters). A literature survey, of analytic, numeric and heuristic techniques for flow lines is given in subsection 3.2.1. For layouts other than flow lines, a literature survey is given in subsection 3.2.2.

A straightforward generalization of the model introduced in subsection 2.3.2 is the N–stage flow line. It is discussed in subsection 3.3.1. Analytic results for three–stage flow lines are discussed in subsection 3.3.2. In subsection 3.3.3 it is shown that three–stage assembly and disassembly systems are essentially equivalent to three–stage flow lines.

Reversibility results are discussed in section 3.4. It is shown that certain networks, which differ in flow direction in certain buffers, are equivalent. Assembly–disassembly networks (subsection 3.4.1) and buffer–sharing networks (3.4.2) are treated.

3.1 NETWORK LAYOUTS AND ANALYSIS TECHNIQUES

Two–stage systems were studied in the previous chapter. In practice, a large variety of system layouts can be encountered. These layouts depend on the way machines and buffers are interlinked. Examples of network layout elements are sketched in Fig 3.1.

The simplest layout structure that can be observed is the *flow line*. In a flow line all products have the same routing over the machines. This situation usually occurs in lines which are dedicated to a single product type or to a limited number of types (often produced in batches).

Other important layout elements are *assembly* and *disassembly processes*. Assembly takes place when a product is assembled out of different parts. If any of these parts is not available, the process stops. In car–manufacturing industries, for example, many car parts are assembled in the final assembly, like the engine, chassis, gear, car interior etc. Think also of process industries, where chemical

substances react to yield a new product. Disassembly can also be observed in chemical industries, when side products emerge in the production.
Even in flow shop–like production systems it may occur that different product types have *different routings* over the capacities. A simple example is the situation of three machines in series, with bypass possibility for products. The first machines operates on type A, the second type B and the third machine is common for all products.
Buffer–sharing can be observed when parallel machines (not necessarily identical) produce the same (or even different) products, which are stored in the same buffer. An example is the insertion department described in section 2.1.
Another important layout element observed in flow shops is the *loop*. A loop in a production system occurs, for example, when a process has failed and some products have to be reprocessed.

The layouts studied in this monograph include flow lines, assembly–disassembly networks and buffer–sharing networks. Assembly and disassembly structures can also be used to model product dependent routings. Assume that the network in Fig 3.1c does not perform disassembly, but produces two product types. Product A requires processing by machine M_1 and product B by machine M_3. About 1/3 of all products is of type A, about 2/3 is of type B. At capacity level, the flow in this system can be modeled continuously, by assuming that machine M_2 performs disassembly. A fraction 1/3 of the product flows to B_1 and a fraction 2/3 flows to B_2. Loops in production systems are left out of scope. In the Chapters 3, 4 and 5 single–product production systems are studied. In later chapters this single product will be interpreted as being an aggregate.

The interesting performance measures may depend on the layout of the system. For example, in assembly systems an important performance measure may be the timely presence of all assembly parts. However, in all layouts, performance measures like throughput, average work–in–process and average throughput times (possibly of different branches) will be important. These are the performance measures which are considered in this text. To obtain such performance measures, the system is modeled by the continuous–flow model introduced in the previous chapter. In principle, there are several possible methods for obtaining these performance measures in the model. The three main techniques are analytic/ numeric methods, simulation and heuristic techniques.

It will be shown in section 3.3, that analytic and numeric methods have only limited application. They can only be applied to analyze systems of up to one or two finite buffers. For larger systems, one of the other techniques has to be used. Simulation is a powerful tool. It can be tailor-made. Simulation of large models has some drawbacks however. One of the drawbacks was mentioned in section 1.4. It is difficult to obtain structural insights in the relations between model parameters by simulation. Another drawback is that, depending on the size of the system, the time needed to obtain a prefixed maximum error in throughput, for instance, may become very long. Results for a simulation on an IBM 4341 computer of an N– machine flow line, with unreliable machines with constant cycle times and finite buffers, are listed in Fig. 3.1.1.

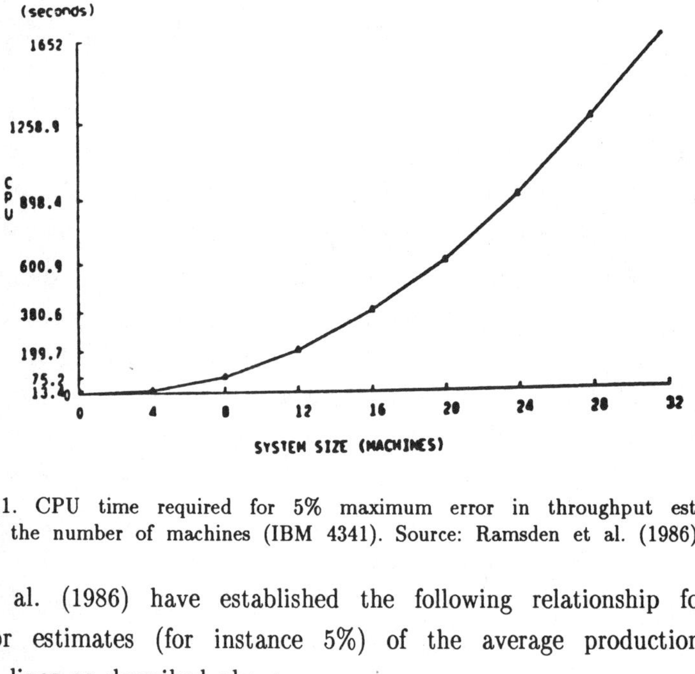

Figure 3.1.1. CPU time required for 5% maximum error in throughput estimates as a function of the number of machines (IBM 4341). Source: Ramsden et al. (1986).

Ramsden et al. (1986) have established the following relationship for constant prefixed error estimates (for instance 5%) of the average production rate, for N–stage flow lines as described above.

$$T(N) = \exp(f(N) \cdot N),$$

where $T(N)$ = CPU time needed for a constant prefixed error estimate of the average production rate, for an N–machine line

N = number of machines

$f(N)$ = non–decreasing function of N with limit $\lim_{N \to \infty} f(N) = f$.

In this text the attention will not be focused on simulation of production systems, but rather on the other two methods. Approximation methods are discussed in Chapters 4 and 5. In this chapter attention is given to analytic and numeric techniques for analyzing production systems with finite buffers and layouts, as mentioned above.

3.2 LITERATURE SURVEY

In this section, the literature on production systems with two or more buffers and three or more stations is reviewed. In section 3.2.1 flow lines are reviewed, in section 3.2.2 more general layouts are discussed.

3.2.1 Literature on N–stage lines

The literature on multistage flow lines can be split up into three parts. The first part consists of analytic and numeric methods, the second of heuristic and the third of simulation methods. The first part splits up again in literature on systems with at most one finite buffer and that on systems with several finite buffers. It appears that analytical results for systems with more than one finite buffer are rare. Throughputs, buffer content distributions etc. can be determined analytically for such systems only in special cases.

Analytic and/or numeric methods can be applied, if buffers are absent or if at most one buffer has *finite* capacity. In that case, exact results can be obtained, even when the system size becomes larger than three machines and two buffers. Exact results can also be obtained when service time distributions are non–exponential. Authors that studied such systems are Hunt (1956), Suzuki (1964), Avi–Itzhak and Yadin (1965), Morse (1965), Buzacott (1968), Barlow and Proschan (1975), Rao (1975a), Prabhu (1967), Shantikumar (1980), Muth and Alkaff (1987) and Schweitzer and Altiok (1988).

Analytic/numeric techniques for lines with three unreliable machines and two finite buffers are given for some special systems by Sheskin (1976), Soyster and Toof (1976) and Soyster et al. (1979). In these fixed–cycle models (see also 2.8.1) the probability of a machine being under repair during a given cycle is independent of the state of that machine during the previous cycle. Gershwin and Schick (1983) and Schick and Gershwin (1978) also studied fixed–cycle, three–stage lines with unreliable machines. The algorithm they use for calculating the throughput of such lines, works for not too large buffer sizes (for large buffers they run into computational difficulties). Sutalaksana and Van Wassenhove (1982) combine analysis and simulation techniques to evaluate a similar three stage model. Comparable results for three–stage continuous–flow lines with unreliable machines and equal production rates were obtained by Coillard and Proth (1984). Newell (1979) obtained results for a three–server fluid model, where the servers have equal service rates, but different variance rates.

For lines consisting of more than three machines, or for more general three–stage lines, the only exploration methods are approximation techniques and simulation. The models for which approximation methods exist can be divided into the same four classes as in section 2.2.

For the first class, discrete–product flow lines with random service times, approximation methods were designed by Harrison (1973a, 1978), Labetoulle and Pujolle (1976), Newell (1979) Boxma and Konheim (1981), Hillier and Boling (1967), Caseau and Pujolle (1979), Altiok (1982, 1986b), Perros and Altiok (1986), Takahashi et al. (1980), Suri and Diehl (1986), Bocharov and Rokhas (1980), Brandwajn and Jow (1986), Kelly (1982), Pollock et al. (1986), Altiok and Ranjan (1988) and Gün and Makowski (1988b).

The second class consists of discrete–product flow–line models with equal cycle times for the machines. Such lines were approximated by Buzacott (1967), Masso and Smith (1974), Gershwin (1987), Dallery et al. (1987).

For continuous–product flow lines (class 3) approximation methods can be found in Glassey and Hong (1986b), De Koster (1987b, 1988), Dallery et al. (1988), Terracol and David (1987a, 1987b) and Ancelin and Semery (1987).

Discrete product models with unreliable machines and random service times (class 4) were studied by Choong and Gershwin (1987), Altiok and Stidham (1983), Klaren (1987) and Gershwin (1988).

Another widely used tool for evaluating flow lines is simulation. See, for instance Barten (1962), Freeman (1964), Anderson and Moodie (1969), Kay (1972), Ramsden et al. (1986) and D'Angelo et al. (1988). Ho et al. (1979, 1983) combined simulation with numeric methods for the evaluation of flow lines.

3.2.2 Literature on general layouts

There is not much literature on production lines with finite capacity buffers and layouts other than flow lines. Nearly all the work in literature concerns discrete product networks, with stochastic (mostly exponential) service times.

Boxma and Konheim (1981) and Altiok and Perros (1986) consider so-called *split* and *merge* configurations. In a split configuration each machine has one or more downstream queues. The destination queue is chosen with a prefixed probability. In a merge configuration, a number of parallel machines, each with its own queue in front of it, supply products to a single queue. The blocking mechanisms studied by these authors differ.

Boxma and Konheim (1981), Altiok and Perros (1987), Perros and Snyder (1986), Takahashi (1988), Jun and Perros (1988) and Lee and Pollock (1988) combined approximation techniques for flow lines with those of split and merge configurations, to obtain approximation methods for the analysis of (nearly) arbitrary networks of finite queues. Boxma and Konheim (1981) give no numeric results. The results of the other authors are good for not too large networks which are reasonably balanced. Kouvatsos and Xenios (1988) give an approximation method based on maximum entropy analysis.

Closed networks are other types of layouts that have been recently studied. Suri and Diehl (1986) study closed networks of finite queues in series, of which, however, the first queue has a capacity larger than the number of products in the system. They give approximations for the throughput and the blocking probabilities. Approximation methods for closed networks with finite-capacity buffers are also given by Akyildiz (1988a), Kouvatsos and Xenios (1988), Dallery and Frein (1988) and Kogan and Krichagina (1988). Exact methods for such networks are studied by Onvural (1988) and Akyildiz (1988b).

There is some literature on assembly-disassembly networks as well. Latouche (1981), Harrison (1973) and Neuts (1981) restrict themselves to a single assembly server which receives independent input streams. Approximation methods of a/d networks were developed by Gershwin (1986a, 1986b), who applied a decomposition method developed for flow lines (Gershwin, 1987) to such networks. The model used by Gershwin (1987) assumes that machines have equal processing times. Liu and Buzacott (1988) give an approximation method for a similar model. MacGregor Smith and Daskalaki (1988) consider buffer-space allocation in automated assembly lines.

In computer science literature, assembly–disassembly is often called *fork/join*, but the idea is the same. A fork occurs when jobs are split up into different parts, each part undergoing a specific number of operations. After these operations the parts may be joined again into a single job. For some literature in this field, mainly dealing with infinite capacity buffers, see Flatto (1985), Flatto and Hahn (1984), Heidelberger and Trivedi (1982, 1983), Nelson and Tantawi (1986) and Liu and Perros (1987).

All the above literature is on discrete–product models. Recently, an approximation method for continuous–flow buffer–sharing networks was developed by Ancelin and Semery (1987).

3.3 N–STAGE LINES

The N–stage continuous flow model is analyzed in this section. In subsection 3.3.1 the model of section 2.3.1 is extended to N machines and N–1 finite intermediate buffers. The analysis given in 3.3.2 shows why even the three–machine system is already difficult to analyze. However, for special cases, e.g. one or more machines being perfect, analytical results can be derived for three–machine lines.

In subsection 3.3.3 attention is paid to three–machine assembly and disassembly networks. There appears to be a close similarity between these systems and three–stage flow lines. It is shown that three–machine assembly and disassembly systems are equivalent to ordinary three–stage flow lines.

3.3.1 The N–stage continuous flow line

The two–stage model of subsection 2.3.2 can be easily extended to N stages. The N–stage flow line is sketched in Fig. 3.3.1.

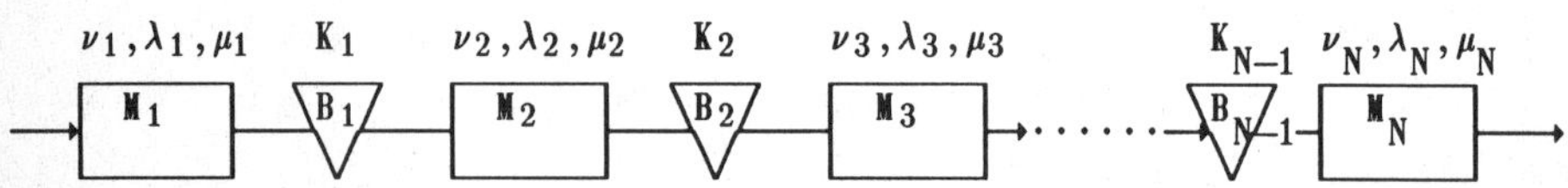

Figure 3.3.1 N–stage flow line with unreliable machines

The product flow in the line in Fig. 3.3.1 is continuous. The production rate of machine M_i is ν_i, its failure rate is λ_i and its repair rate is μ_i. Both lifetimes and repair times are exponentially distributed. The capacity of buffer B_i is K_i (i=1,.., N–1). If buffer B_i is full, then machine M_i will have to slow down to the production rate at which M_{i+1} is working, if this rate is lower than the actual production rate of M_i. Hence, if B_i is full and M_{i+1} is down, then M_i is blocked. Similarly, M_i may be slowed down or starved by an empty buffer B_{i-1} (i=2,..,N). Machine M_1 is never slowed down by lack of input and M_N is never slowed down by lack of room to store its output. It is assumed that the failures of all machines are time-dependent. The throughput of this line is denoted by v(**K**), where $\mathbf{K}=(K_1,..,K_{N-1})^T$.

3.3.2 Analysis of three-stage lines

For computational reasons, the flow line analysis is carried out for three-stage lines only. It is largely based on De Koster and Wijngaard (1986, 1987b). The state of the three-stage line is described completely by the triple (i,x,y), where i is the machine-state of the system, as given in Table 3.3.1, and the pair (x,y) denotes the inventory levels in B_1 and B_2, respectively.

M \ state	1	2	3	4	5	6	7	8
1	1	0	0	1	1	0	0	1
2	0	1	0	1	0	1	0	1
3	0	0	1	0	1	1	0	1

Table 3.3.1. Possible machine states. A one in the row of machine M_i means M_i is up, a zero means M_i is down because of a failure.

As in section 2.3.3 regeneration points can be distinguished and the "cost" per cycle can be calculated. Let

$\alpha_i(x,y)$ be the cost rate in machine state i (see Table 3.3.1) with inventory level x in B_1 and y in B_2, $(x,y) \in [0,K_1]\times[0,K_2]$, and let

$f_i(x,y)$ be the expected cost till the end of the cycle, if we are now in machine state i, with inventory levels x and y, respectively.

The expected cost per cycle C_T can be expressed in the α_i's and f_i's, depending on the chosen regeneration point. For instance, if the regeneration point is chosen to be (6,0,0), then

$$C_T = \frac{\alpha_6(0,0)}{\mu_1+\lambda_2+\lambda_3} + \frac{\mu_1}{\mu_1+\lambda_2+\lambda_3}f_8(0,0) + \frac{\lambda_2}{\mu_1+\lambda_2+\lambda_3}f_3(0,0) + \frac{\lambda_3}{\mu_1+\lambda_2+\lambda_3}f_2(0,0) \qquad (3.1)$$

In (3.1), $\frac{1}{\mu_1+\lambda_2+\lambda_3}$ is the expected duration of a stay in machine state 6 and, for example, $\frac{\mu_1}{\mu_1+\lambda_2+\lambda_3}$ is the conditional probability of a transition to machine state 8. The choice of the regeneration point is important, since the corresponding state has to be positive recurrent. The state (6,0,0) can only be a regeneration point if $\lambda_1 \neq 0$.

By distinguishing between the cost in a certain small time interval Δ and the cost during the rest of the cycle, a system of partial differential equations for the f_i's can be derived. For f_1 we obtain, for $(x,y)\in[0,K_1)\times[0,K_2]$:

$$\begin{aligned} f_1(x,y) = \ & \alpha_1(x,y)\Delta + (1-\lambda_1\Delta)(1-\mu_2\Delta)(1-\mu_3\Delta)f_1(x+\nu_1\Delta,y) + \\ & \lambda_1\Delta(1-\mu_2\Delta)(1-\mu_3\Delta)f_7(x+o(\Delta),y) + \\ & (1-\lambda_1\Delta)\mu_2\Delta(1-\mu_3\Delta)f_4(x+o(\Delta),y+o(\Delta)) + \\ & (1-\lambda_1\Delta)(1-\mu_2\Delta)\mu_3\Delta\ f_5(x+o(\Delta),y+o(\Delta)) + o(\Delta) \end{aligned}$$

Dividing by Δ and letting $\Delta \to 0$ we obtain, for $x\in[0,K_1)$, $y\in[0,K_2]$

$$-\nu_1 f_{1x}(x,y) = \alpha_1(x,y) - (\lambda_1+\mu_2+\mu_3)f_1(x,y) + \lambda_1 f_7(x,y) + \mu_2 f_4(x,y) + \mu_3 f_5(x,y) \qquad (3.2)$$

where $f_{1x}(x,y)$ is shorthand for $\frac{\partial f_1}{\partial x}(x,y)$.

The equations for the other machine states can be derived similarly.
In matrix notation the system of partial differential equations can be rewritten as

$$-\mathbf{A}\mathbf{f}_x(x,y) - \mathbf{B}\mathbf{f}_y(x,y) = \boldsymbol{\alpha}(x,y) + \mathbf{C}\mathbf{f}(x,y), \qquad (3.3)$$

where $\mathbf{f}(x,y)$ is the vector $(f_1(x,y),..,f_8(x,y))^T$, $\boldsymbol{\alpha}(x,y)$ is the vector $(\alpha_1(x,y),..,\alpha_8(x,y))^T$, $\mathbf{A}$ is the drift matrix of buffer B_1, namely $\mathrm{Diag}(\nu_1,-\nu_2,0,\nu_1-\nu_2,\nu_1,-\nu_2,0,\ \nu_1-\nu_2)$, $\mathbf{B}$ is the drift matrix of buffer B_2, namely $\mathrm{Diag}(0,\nu_2,-\nu_3,\nu_2,-\nu_3,\nu_2-\nu_3,0,\nu_2-\nu_3)$ and

$$
C = \begin{bmatrix}
-\lambda_1-\mu_2-\mu_3 & 0 & 0 & \mu_2 & \mu_3 & 0 & \lambda_1 & 0 \\
0 & -\lambda_2-\mu_1-\mu_3 & 0 & \mu_1 & 0 & \mu_3 & \lambda_2 & 0 \\
0 & 0 & -\lambda_3-\mu_1-\mu_2 & 0 & \mu_1 & \mu_2 & \lambda_3 & 0 \\
\lambda_2 & \lambda_1 & 0 & -\lambda_1-\lambda_2-\mu_3 & 0 & 0 & 0 & \mu_3 \\
\lambda_3 & 0 & \lambda_1 & 0 & -\lambda_1-\mu_2-\lambda_3 & 0 & 0 & \mu_2 \\
0 & \lambda_3 & \lambda_2 & 0 & 0 & -\mu_1-\lambda_2-\lambda_3 & 0 & \mu_1 \\
\mu_1 & \mu_2 & \mu_3 & 0 & 0 & 0 & -\mu_1-\mu_2-\mu_3 & 0 \\
0 & 0 & 0 & \lambda_3 & \lambda_2 & \lambda_1 & 0 & -\lambda_1-\lambda_2-\lambda_3
\end{bmatrix}
$$

The dimension of the system (3.3) can be reduced by one, by substituting the seventh equation into the other ones.

The boundary conditions corresponding to a differential equation, can be derived by considering the boundaries for which the particular equation does not hold. These boundaries have to be split up into *sides* of the rectangle $[0,K_1]\times[0,K_2]$ and *corner points.* In such sides and corner points, machines are slowed down, blocked or starved and therefore have to adapt their speeds. This means that a boundary condition can be obtained from the corresponding differential equation by changing the speeds of one or more machines. We derive the boundary conditions corresponding to the fourth equation to provide an example.

For the rectangle side with $x\in[0,K_1)$, $y=K_2$ we have

$$
-\nu_1 f_{4x}(x,K_2) = \alpha_4(x,K_2) - (\lambda_1+\lambda_2+\mu_3)f_4(x,K_2) + \lambda_1 f_2(x,K_2) + \lambda_2 f_1(x,K_2) + \\ + \mu_3 f_8(x,K_2)
$$

For the corner point with $x=K_1$, $y=K_2$ we have

$$
0 = \alpha_1(K_1,K_2) - (\lambda_1+\lambda_2+\mu_3)f_4(K_1,K_2) + \lambda_1 f_2(K_1,K_2) + \lambda_2 f_1(K_1,K_2) + \\ + \mu_3 f_8(K_1,K_2)
$$

Furthermore, for $\nu_2 \geq \nu_1$, $x=0$, $y\in[0,K_2)$ we have

$$
-\nu_1 f_{4y}(0,y) = \alpha_4(0,y) - (\lambda_1+\lambda_2+\mu_3)f_4(0,y) + \lambda_1 f_2(0,y) + \lambda_2 f_1(0,y) + \mu_3 f_8(0,y)
$$

For $\nu_2 \leq \nu_1$, $x=K_1$, $y\in[0,K_2)$ we have

$$
-\nu_2 f_{4y}(K_1,y) = \alpha_4(K_1,y) - (\lambda_1+\lambda_2+\mu_3)f_4(K_1,y) + \lambda_1 f_2(K_1,y) + \lambda_2 f_1(K_1,y) + \\ + \mu_3 f_8(K_1,y)
$$

However, depending on the chose regeneration point, one of these boundary conditions may have to be changed for the condition that the corresponding f_i has to be continuous at the regeneration point.

When all machines are unreliable, analytic solutions of the system (3.3) can be obtained only, if all machine speeds are equal. Coillard and Proth (1984) obtained an analytic solution for the steady state probabilities of system states for the case with all machines equal, and equal buffer capacities. In this case, the solution of the homogeneous system can be conjectured and the constants in it can be evaluated using the boundary conditions. For the general case, with all machines unreliable and all machine speeds different, it is far from straightforward to obtain explicit expressions for the cycle cost (3.1). The difficulties are largely due to the complicated boundary conditions. The constants in a conjectured general solution of the homogeneous system have to be evaluated using these boundary conditions. Therefore, the general case has only been solved via numeric methods. In the following subsections some special cases will be discussed, in which one or more machines are perfectly reliable. In these cases the system can be solved easily.

3.3.2.1 Two machines perfect, one machine unreliable

Assume that two machines are perfectly reliable in the three–stage line.

The solution is trivial if the production rate of the unreliable machine is the lowest. The throughput then equals $\min_i\{\mathrm{eff}_i\}$, where eff_i is the net production rate of M_i. This can be easily understood, since the unreliable machine will never be slowed down or forced down.

If the production rate of the unreliable machine is not the lowest, the determination of the line throughput is not trivial in all cases. We consider three different cases.

1. M_2 unreliable
2. M_1 unreliable
3. M_3 unreliable.

1. Only M_2 unreliable.

– If $\nu_2 > \min\{\nu_1,\nu_3\}$ and $\nu_2 \leq \max\{\nu_1,\nu_3\}$, then the line is easily analyzed. For instance, if $\nu_1<\nu_2\leq\nu_3$, then B_2 will become empty after some time and remain empty. Therefore the throughput of this line equals the throughput of the two–

stage line M_1–B_1–M_2. The analysis of such a line is given in appendix A.1.

– For the case $\nu_2 > \max\{\nu_1, \nu_3\}$ we distinguish two subcases.
a) $\nu_1 = \nu_3$
b) $\nu_1 > \nu_3$ (the case $\nu_1 < \nu_3$ is similar)

a) The case $\nu_1 = \nu_3$
The system of partial differential equations can be solved (see De Koster and Wijngaard, 1986). It appears that the throughput of the line equals the throughput of the two-stage line M_1–B–M_2, where B is a buffer of capacity $K=\min\{K_1,K_2\}$. This can also be seen as follows. Let a(t) = state of M_2 at time t.

Suppose a(0)=0 and the buffer contents at time t=0 of B_1 is K_1 and of B_2 is 0. Let $\{(t,a(t)) \mid t \in \mathbb{R}_+\}$ be a stochastic realization of M_2. The patterns of the contents of B_1 and B_2, corresponding to such a realization are sketched in Fig. 3.3.2, for $K_1>K_2$. A similar picture can be drawn if $K_1 \leq K_2$.

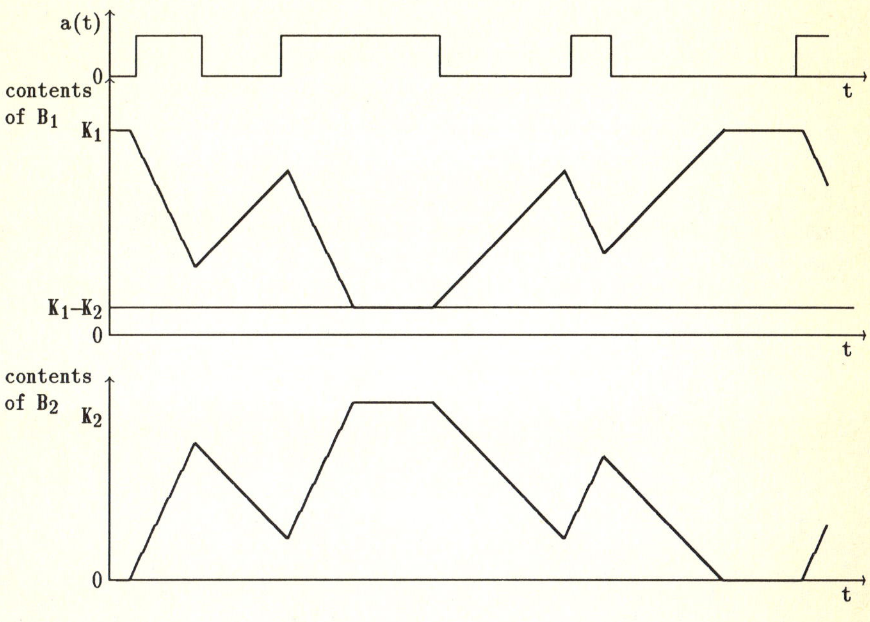

Figure 3.3.2. Behavior of the contents of B_1 and B_2, in case $K_1>K_2$

The contents of B_1 and B_2 are negatively correlated: if the content of B_2, X_2, equals x_2, then $X_1=K_1-x_2$. If $K_1>K_2$, then B_1 will never become empty any more, after is has been full, and if $K_1=K_2$, then B_1 is full if and only if B_2 is empty. In both cases, the graph of the contents of B_2 in this line is identical to the graph of the content of B_2 in the two-stage line M_2–B_2–M_3 and hence the throughputs are also equal. A similar observation can be made if $K_1 \leq K_2$.

Another remarkable conclusion can be drawn from the case $K_1=K_2$, for the two lines M_1–B_1–M_2 and M_2–B_1–M_1 (note that the machines M_3 and M_1 are identical):

1. the throughputs are equal and furthermore
2. the behavior of the content of B_1 in the first two-stage line, $X_1(t)$, is exactly the opposite of the behavior of the content of B_1 in the second line, X_1'. That is $X_1'(t) = K_1 - X_1(t)$, for all stochastic realizations of the failure and repair process of M_2 and for all points in time, provided $X_1'(0) = K_1 - X_1(0)$. Hence also the mean buffer contents will be reversed: $EX_1' = K_1 - EX_1$.

For an arbitrary two-stage line M_1–B_1–M_2, the two-stage line M_2–B_1–M_1 is said to be the *reverse* of the line, or equivalently, it arises from the line by *flow reversal.* In section 3.4 it will be shown that properties like the ones mentioned above hold for more general production systems.

b) The case $\nu_1 > \nu_3$

In this case the throughput only be calculated in general for the case $\frac{K_1}{\nu_2-\nu_1} \geq \frac{K_2}{\nu_2-\nu_3}$. In the latter case, it can be proven that M_2 is never slowed down to rate ν_1 and hence the line is equivalent with M_2–B_2–M_3. The quantity $\frac{K_1}{\nu_2-\nu_1}$ is the time needed to empty the full buffer B_1, if there are no breakdowns, and $\frac{K_2}{\nu_2-\nu_3}$ is the time needed to fill the empty buffer B_2. Starting with a full buffer B_1 and an empty buffer B_2, the second buffer will always be full before the first one is empty (in case of an inequality). In case of an equality it can occur that the first buffer is empty, but only if the second buffer is full simultaneously.

This also becomes clear from the sketch of the possible realizations of the buffer contents (X_1,X_2) in Fig. 3.3.3. The two arrows in each point in Fig 3.3.3 indicate in which direction the buffer contents move if M_2 is either up or down. Define S as the convex hull of $(K_1,0)$, (K_1,K_2) and $(-\frac{\nu_2-\nu_1}{\nu_2-\nu_3}K_2+K_1,K_2)$ (the shaded area in Fig. 3.3.3).

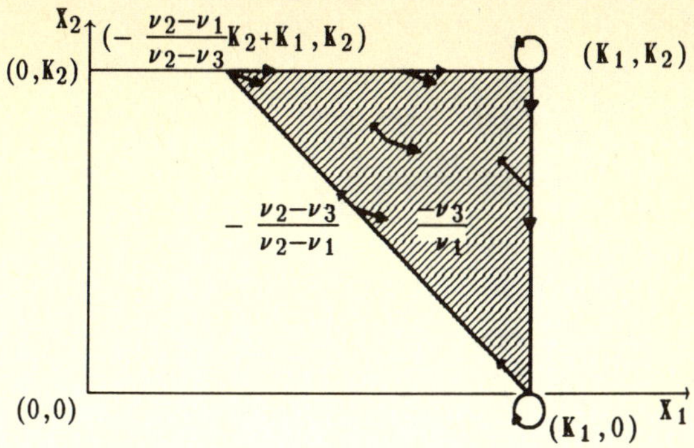

Fig. 3.3.3. Possible realizations for (X_1,X_2), in case $\frac{K_1}{\nu_2-\nu_1} > \frac{K_2}{\nu_2-\nu_3}$

Assume we start at time t=0 with M_2 down and $(X_1,X_2)=(K_1,0)$. As long as M_2 stays down we have $(X_1,X_2) = (K_1,0)$. However, when M_2 is repaired X_1 decreases: the buffer contents will move along the line

$X_2 =-\frac{\nu_2-\nu_3}{\nu_2-\nu_1}(X_1-K_1)$, until $X_2=K_2$ or M_2 breaks down. This line intersects the line $X_2=K_2$ in the point $(-\frac{\nu_2-\nu_1}{\nu_2-\nu_3}K_2+K_1,K_2)$, with abscis > 0. It can be seen easily that if at time t $(X_1(t),X_2(t)) \in S$, we have $(X_1(t+\Delta),X_2(t+\Delta)) \in S$, for all $\Delta \geq 0$. Therefore $X_1(t) > 0$ for all $t \in \mathbb{R}_+$.

If $\frac{K_1}{\nu_2-\nu_1} = \frac{K_2}{\nu_2-\nu_3}$ it can occur that $X_1(t) = 0$. However, this occurs only if $X_2(t) = K_2$. Since $\nu_1 > \nu_3$ we see that also in this case the line is equivalent with the line M_2–B_2–M_3.

2. Only M_1 unreliable.

The following subcases are considered

a) $\nu_2 > \nu_1 > \nu_3$

b) $\nu_1 > \nu_2 > \nu_3$

a) The case $\nu_2>\nu_1>\nu_3$

This system can be analyzed by considering the aggregate inventory level $Y=X_1+X_2$. This is because M_1 has to slow down (to rate ν_3) if and only if $Y=K_1+K_2$ and M_3 is starved if and only if M_1 is down and $Y=0$. The line will behave in the same way as M_1–B–M_3, where B is a buffer of capacity $K=K_1+K_2$.

b) The case $\nu_1>\nu_2>\nu_3$

The line reduces to a two–stage line only if $\frac{K_1}{\nu_1-\nu_2} \geq \frac{K_2}{\nu_2-\nu_3}$. This condition means that B_2 is filled before B_1 is filled. In this case, the buffer contents (X_1,X_2) will move in the shaded area and the rectangle side $[0,K_1]\times K_2$, indicated in Fig. 3.3.4.

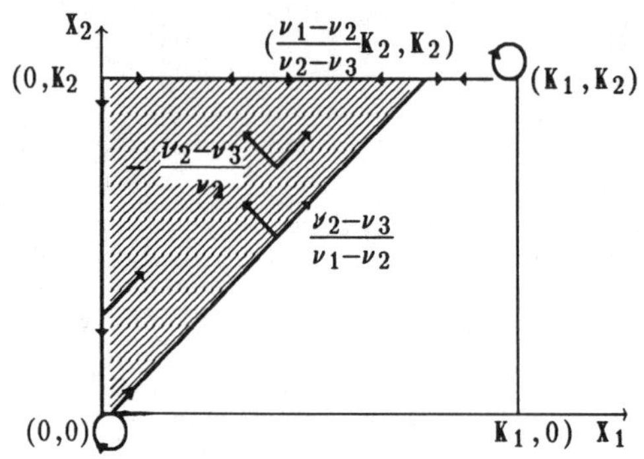

Fig. 3.3.4. Possible realizations for (X_1,X_2), in case $\frac{K_1}{\nu_1-\nu_2} > \frac{K_2}{\nu_2-\nu_3}$

If $\frac{K_1}{\nu_1-\nu_2} \geq \frac{K_2}{\nu_2-\nu_3}$ it can be seen from Fig 3.3.4, that M_1 is slowed down only (to rate ν_3) if the aggregate buffer stock $Y=K_1+K_2$ (because B_2 is full before B_1 is full) and M_3 can only be starved if $Y=0$ (because B_2 is empty before B_1 is empty). Hence the system is equivalent with M_1–B–M_3, with B a buffer of capacity $K=K_1+K_2$.

3. Only M_3 unreliable.

In section 3.4 it will be shown that by reversing the flow direction in this line, a sample path equivalent line is obtained. Hence, this case is equivalent to the previous one.

3.3.2.2 More than one machine unreliable

Lines with only one perfect machine are in general difficult to analyze. There are only a few cases where the line is reducible to a two–stage line.

- If M_1 is perfect and $\nu_1 \geq \nu_2$, then the line has the same throughput as the line M_2–B_2–M_3.
- If M_2 is perfect and $\nu_2 \geq \max\{\nu_1,\nu_3\}$, then M_1 is slowed down or blocked only,

if the aggregate buffer $Y=K_1+K_2$. M_3 can only be slowed down or starved if $Y=0$. Hence the system has the same throughput as the line M_1–B–M_3, where B is a buffer of capacity $K=K_1+K_2$.

3.3.3 Three–machine assembly–disassembly networks

The analysis of subsection 3.3.2 can be easily extended to three–machine assembly–disassembly networks. A three–machine assembly network is sketched in Fig. 3.1b. A three–machine disassembly network is sketched in Fig. 3.1c. In the network in Fig. 3.1b, machine M_2 takes one unit of product from B_1 and one from B_2, and assembles them into a unit of a new product. Machine M_2 is slowed down in this network if at least one of the upstream buffers is empty. In the network sketched in Fig. 3.1c, M_2 performs disassembly operations. It is slowed down if either of the buffers B_1 or B_2 is full. The product flow in the models is again assumed to be continuous. The machines have speeds ν_i, failure rates λ_i and repair rates μ_i, just as in subsection 3.3.2.

The analysis of the networks in Fig. 3.1b and Fig. 3.1c is similar to that in subsection 3.3.2. This can intuitively be understood as follows. For the assembly network in Fig. 3.1b, for example, we can compare the behavior of the network, starting with B_1 empty and B_2 full, with the three–stage flow line M_1–B_1–M_2–B_2–M_3, starting with both B_1 and B_2 empty. In both systems, machine M_3 will be slowed down to the minimum rate of M_1 and M_2. In both the assembly network and the flow line, B_1 fills or empties at the same rate. If buffer B_1 will be full in the assembly network, then it will also be full in the flow line. In both networks M_1 is then slowed down to the rate of M_2. Furthermore, if B_2 in the flow line increases at rate $\nu_2-\nu_3$, then it decreases in the assembly network at the same rate. Hence, if B_2 is full in the flow line, then B_2 is empty in the assembly network and, in either case, M_2 is slowed down to the rate of M_3. A similar reasoning holds for the disassembly network in Fig. 3.1c, compared with the flow line M_1–B_1–M_2–B_2–M_3. This implies that the analysis of three–stage assembly and disassembly networks is similar to that of three–stage flow lines.

This intuitive reasoning can be made formal for the networks in Figs. 3.1b and 3.1c, by defining machine states as in Table 3.3.1 and α_i's and f_i's as in sub–section 3.3.2. The point $(6,0,K_2)$ is chosen as regeneration point for the network in Fig. 3.1b, and $(6,K_1,0)$ is chosen for the network in Fig. 3.1c. It is now found that the systems of partial differential equations, the boundary equations and the equations for the expected cycle cost for the system in Fig. 3.1b and the flow line M_1–B_1– M_2–B_2–M_3, are related. The equations for the assembly network can be obtained from the system of equations (3.3) and corresponding boundary equations and from (3.1), respectively, by applying the mapping

$$(x,y) \mapsto (x,K_2-y) \tag{3.4}$$

This implies that the expected costs per cycle of these systems are equal. Hence the cycle time, throughput, mean buffer content of B_1 etc. are equal, but the mean buffer content of B_2 in Fig. 3.1b, EX_2', equals K_2-EX_2, with EX_2 the mean buffer content of B_2 in the flow line, described in section 3.3.2.

A similar conclusion can be drawn for the network in Fig. 3.1c, compared with the system in 3.3.2. The equations can be obtained from each other by application of the mapping

$$(x,y) \mapsto (K_1-x,y) \tag{3.5}$$

And hence the throughputs and mean buffer contents of B_2 are equal. The mean buffer content of B_1, EX_1', equals K_1-EX_1, where EX_1 is the mean content of B_1 in the flow line.

The networks in Figs. 3.1b and 3.1c could also have been compared with the flow line M_3–B_2–M_2–B_1–M_1. It would have been found that again the equations for this flow line could have been obtained by applying a mapping $(x,y)\mapsto(K_1-x,y)$ (to equations for the network in Fig. 3.1b) or $(x,y)\mapsto(x,K_2-y)$ (network in Fig. 3.1c).

In view of the correspondence in buffer behavior in these networks, we call two different networks, consisting of the same machines and (finite) buffers and with the same layouts, except for flow direction, *sample–path–equivalent* if the following condition is satisfied.

> Assume that for every buffer with capacity K the property holds that the content is X(0) in the one network at time 0, and the content is either K–X(0) or X(0) in the other, depending on whether the flow direction in that buffer is reversed or. Let the content of the buffer in the one network be X(t) at time t, then the content in the other network is K–X(t) or X(t), depending on whether or not the flow direction in the buffer is reversed.

For the particular networks studied, this condition implies that the throughputs in sample-path-equivalent networks are also equal. With this definition we can conclude that the networks sketched in Fig. 3.3.5 are sample-path-equivalent. The equivalence relationships are indicated in Fig. 3.3.5.

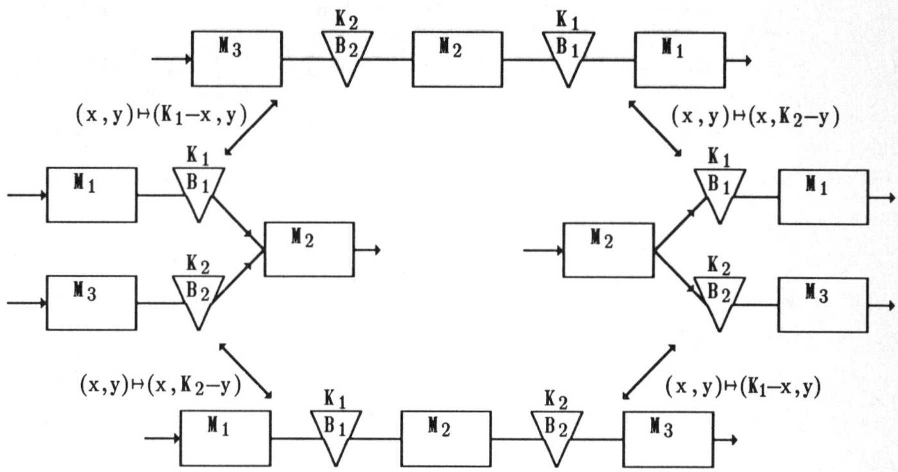

Figure 3.3.5. Four equivalent networks

A network is called *sample-path-reversible*, if each network arising from it by flow reversal in one or more of the buffers, under addition of extra input or output buffers if necessary, is sample-path-equivalent to the original network.

In the next section these properties will be shown to hold more generally. Section 3.4 deals with equivalence properties of continuous-flow production systems. Such network equivalencies are often convenient in network analysis. For instance, calculating the output behavior of a certain production system may be equivalent to calculating the input behavior of the reversed system. In this way computation and simulation time can be saved. Furthermore it can provide a proper insight into the behavior of certain production networks. Readers not interested in these equivalence properties may skip section 3.4 and proceed with Chapter 4.

3.4 REVERSIBILITY OF PRODUCTION NETWORKS

Several authors have paid attention to the reversibility of queueing systems with finite buffers. The term *reversibility* usually denotes that reversing the complete network, that is, reversing the direction of the flow in *each* buffer, leaves certain performance measures (usually the throughput) invariant in certain networks. Nearly all authors study discrete product networks. Muth (1979) studies flow lines with finite intermediate buffers and work stations with stochastic service times. He proved that the throughput of such a line remains unchanged under the reversal of flow. Yamazaki et al. (1985) proved that throughput invariance on flow reversal also holds for flow lines with parallel work stations with deterministic service times. For flow lines with single stochastic work stations and finite buffers Yamazaki et al. (1985, 1978) proved that on flow reversal the distributions of the time of the n–th departure epochs are also identical, assuming that the systems are empty initially. Melamed (1986) studied single buffer systems with parallel work stations. In all these references the blocking mechanism is the same. If a work station finishes an operation and the succeeding buffer is full, then the work station becomes blocked and cannot serve other items (*transfer blocking*, see 2.2).

Ammar (1980) and Ammar and Gershwin(1981) studied *communication blocking*: a work station does not start an operation if the succeeding buffer is filled completely. They considered networks of work stations where assembly/disassembly (or shortly, a/d) of products takes place. The layout of these networks is such that a buffer is linked to exactly one upstream and one downstream machine, but a machine may have several upstream or downstream buffers (see Figs. 3.1b and 3.1c). If a work station has more than one upstream buffer it performs the assembly of one unit of each of the different buffers. This work station can only work if all upstream buffers are nonempty. If a work station has more than one downstream buffer, it performs the disassembly of the product into one unit for each of the buffers. It can only operate if all downstream buffers are nonfull. Ammar and Gershwin (1981) demonstrated similar properties as sample–path–reversibility for discrete product a/d networks.

Reversibility properties of two–stage continuous–flow lines were studied by Hontelez (1985).

In this section, reversibility properties of production networks with continuous product flow will be discussed. Two types of networks with different layouts are studied. In subsection 3.4.1 continuous–flow assembly–disassembly networks are discussed. It is shown that such networks are sample–path–reversible. The second type of layouts studied (subsection 3.4.2), are layouts in which machines share common upstream or downstream buffers. In such buffer–sharing networks it is also possible to reverse the direction of flow through a single buffer *b* resulting in a sample–path–equivalent network. However, this equivalent network is not necessarily a buffer–sharing network any longer. In general, the resulting network is a combination of a buffer–sharing and an a/d network. Only if the flow through *all* buffers is reversed, is a buffer–sharing network obtained.

To demonstrate sample–path–reversibility for a/d networks and sample–path–equivalence between buffer–sharing networks, two propositions will be proven in subsections 3.4.1 and 3.4.2, respectively. The results in the next subsections are based on Corten and De Koster (1987). Quantities denoted by a prime will in the sequel refer to the reversed network.

3.4.1 Reversibility in assembly–disassembly networks

In this section, assembly–disassembly networks (or shortly, a/d networks) are considered. That is, it is assumed that machines may have multiple upstream or downstream buffers, but each buffer has, at most, one upstream machine and one downstream machine. The production flow in the network is continuous. A sample network is sketched in Fig. 3.4.1. All buffers in Fig. 3.4.1, except B_1, B_2, B_{18} and B_{19} are supposed to have finite capacity. The capacity of buffer j is denoted by K_j.

Buffers not preceded by machines are called *source buffers.* Source buffers are never empty. Buffers not followed by machines are called *sink buffers.* Sink buffers are never full. These buffers are not really necessary, but by introducing them, each machine has at least one upstream and one downstream buffer. In this section, the set of machines in a network is denoted by *M*, the set of buffers by *B*.

Machines that have more than one upstream buffer perform an assembly operation. That is, they assemble one unit of product by taking a fixed amount of product out of each buffer. We assume one unit of product is taken out of each upstream buffer, but different assembly ratios do not complicate the proof of proposition 3.1, in which the reversibility of a/d networks is demonstrated. A machine that has more than one downstream machine disassembles a unit of product into one unit of each disassembly part. Equal disassembly ratios are also not really essential either. A machine that performs an assembly and/or a disassembly operation is called an assembly–disassembly machine (a/d machine).

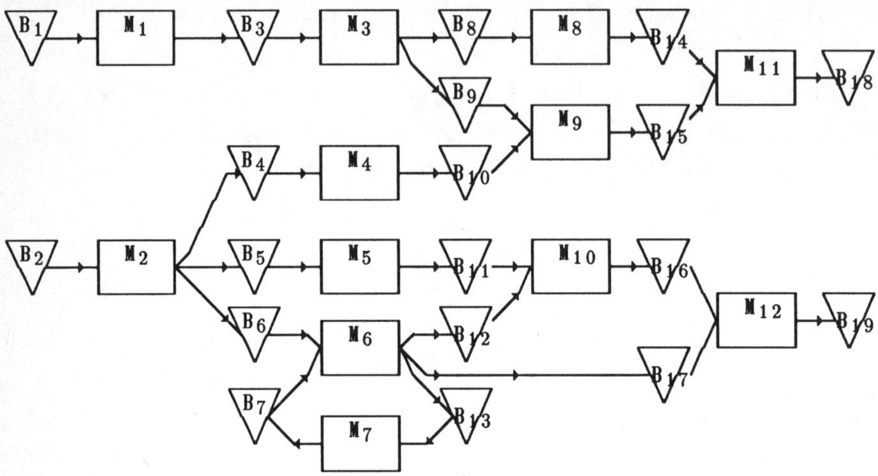

Fig. 3.4.1 Assembly–disassembly network

Due to the finiteness of the buffer capacities, machines in a/d networks may be slowed down. An a/d machine may be slowed down whenever only one of its upstream buffers is empty or only one of this downstream buffers is full. Assume, for instance, that in the network in Fig. 3.4.1 machine M_{11} can work at speed v_{11}, machine M_8 at speed v_8, B_{14} is empty and B_8 and B_{15} are not empty. If $v_8 < v_{11}$, then M_{11} is slowed down to rate v_8. Since all machines are connected by buffers, slowing down of machines may proliferate through the network. A machine may be slowed down through full downstream or empty upstream buffers. Assume that a machine is followed by a full buffer. Then this machine may be slowed down by the downstream machine of that buffer. This last machine may in turn be slowed down through another empty upstream, or a full downstream buffer, etc. We say that the slowing down proliferation takes place via so–called *proliferating paths.* A machine is on a proliferating path from machine i, at time t, if this machine can slow down machine i, through a chain of full or empty buffers (if its production rate were smaller than that of machine i). Formally the following definition can be given.

We say that machine j in M is on a *proliferating path from machine i at time t*, if there exists a directed path in the network from machine i to machine j at time t, so that for every buffer on the path it holds that if the path goes upstream through the buffer, the buffer is empty and if the path goes downstream through the buffer, then it is full. See Fig. 3.4.2.

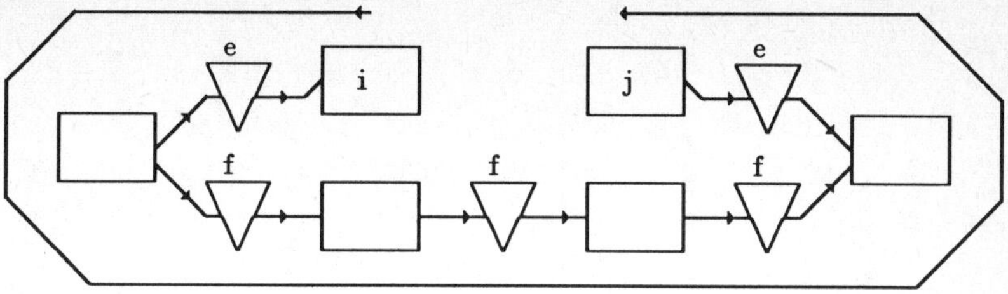

Figure 3.4.2. Proliferating path from machine i to machine j. All buffers are either empty (e) or full (f).

Note that proliferating paths are variable over time. Furthermore they have a direction: machine j in Fig. 3.4.2 can slow down machine i, but machine i cannot slow down machine j.
For i $\in M$, $R_i(t)$ is the set of all machines which are on a proliferating path from machine i at time t (including i).
Knowledge of $R_i(t)$ is necessary, to determine the speed of machine i at time t.

In proposition 3.1 it is shown that a/d networks are sample–path–reversible. In order to prove this proposition, it is necessary to make some assumptions on the behavior of the machines.

It is assumed that each machine can be in countably many states {0,1,2,..}. The stochastic variable representing the machine state of machine $i \in M$, is denoted by $a_i(t)$. A constant machine speed is associated with each machine state (as in section 2.4). The sojourn time in each state is a random variable (not necessarily exponentially distributed). After a sojourn time in a state, a transition is made to another state. This transition is determined by an irreducible Markov transition matrix. Two types of *change points* are distinguished:

- *change points in machine states*: points in time at which a machine changes its state. It is assumed that these change points do not condense
- *buffer filling or emptying points*: points in time when buffers become full or empty.

It is assumed that these two types of change points do not coincide. Note, that the machine speeds in a network only change at such change points.

The aim is now to define the behavior of $a_i(t)$ so, that both time–dependent and state–dependent transitions are covered. If the transition rates are time–dependent, they do not depend on the time the machine has been slowed down through full or empty buffers. If a machine has state–dependent transition rates, the transition rates may be influenced by the amount and the time duration of the slow down. As an example, consider an unreliable machine, with production rate 6, (exponential) failure rate 0.1 and (exponential) repair rate. This machine has state–dependent failure behavior of the type as modeled in subsection 2.3.6. Assume that we are now at time point 0 with the machine up. The failure time is determined by a random draw from an exponential distribution with rate 0.1. Assume that the failure time is 10. Assume furthermore that the machine is slowed down to rate 0 at time 5 and at time 7 it is slowed down to rate 2. Arriving at t=5, due to the memoryless property of the exponential distribution, the new failure time is determined by a random draw from an exponential distribution with rate $0 \times 0.1/6=0$. At time t=7, the new failure time is determined by a random draw from an exponential distribution with rate $2 \times 0.1/6$.

More general,
let the state of machine i at change point t be $\{a_i(t)\}$. The sojourn time in that state is determined as follows. If the machine is not slowed down at time t, then the sojourn time is determined by a random draw from the same distribution as if the machine were in stand alone position. If the machine *is* slowed down at time t, then the sojourn time is determined by a random draw from a distribution that may depend on the amount and duration of slow down. In all cases, the sojourn time distribution is assumed to have a positive mean. (3.6)
These change points generated by a random draw from a certain distribution are denoted by *projected change points*, since these generated change points are not necessarily realized.

The *stand–alone* (or *potential*) machine speed, corresponding to the realization of the machine state $\{(t,a_i(t));\ t \geq 0\}$ is denoted by $v_i(t)$. The buffer content of buffer $j \in B$ at time t is denoted by $X_j(t)$. The real machine speed of machine i, as influenced by the network is denoted by $w_i(t)$. This machine speed is given by

$$w_i(t) = \min\{v_j(t);\ j \in R_i(t)\} \tag{3.7}$$

Proposition 3.1 (sample-path-reversibility of a/d networks).

Let N be an a/d network, with machines modeled as above, with sojourn time distributions in each state as defined above. Let $\ell \in B$, the set of buffers in N, and let N' be the network arising from N by reversing the direction of the flow in buffer ℓ. If this results in a machine without upstream buffer, then add a source buffer supplying this machine. If the flow reversal results in a machine without downstream buffer, then add a sink buffer obtaining parts from that machine. The cardinality of M is denoted by n, the cardinality of B by m.

Let $\{(t,a_1(t),..,a_n(t));\ t\geq 0\}$ be a stochastic realization of the machine states in N, $\{(t,X_1(t),..,X_m(t));\ t\geq 0\}$ the corresponding realization of the buffer contents and $\{(t,w_1(t),..,w_n(t));\ t\geq 0\}$ the corresponding realization of the machine speeds in N.

Define $\{X_j'(0)\} = \{X_j(0)\}$, $j\in B$, $j\neq\ell$, $\{X_\ell'(0)\} = \{K_\ell - X_\ell(0)\}$.

$\{a_i'(0)\} = \{a_i(0)\}$, $i\in M$

Then,

$\{(t,a_1(t),..,a_n(t));\ t\geq 0\}$ is a stochastic realization of the machine states in N', $\{(t,X_1(t),..,K_\ell - X_\ell(t),..,X_m(t));\ t\geq 0\}$ the corresponding realization of the buffer contents and $\{(t,w_1(t),..,w_n(t));\ t\geq 0\}$ the corresponding realization of the machine speeds N'.

Proof. Note first that, if at time t we have $\{a_i'(t)\}=\{a_i(t)\}$ for all $i\in M$ and $X_j'(t)=X_j(t)$, for all $j\neq\ell$, $X_\ell'(t)=K_\ell-X_\ell(t)$, then for all $i\in M$, the set of machines that are on a proliferating path from machine i at time t, is the same in N and N'. Note that the flow direction through buffer ℓ is reversed, but simultaneously $X_\ell'(t)=K_\ell-X_\ell(t)$ and hence if machine j slows down machine i by a proliferating path through buffer ℓ in N, then it also slows down machine i in N'. This means that

$$R_i'(t)=R_i(t) \text{ and hence, by (3.7), } w_i'(t)=w_i(t), \text{ for all } i\in M. \tag{3.8}$$

The proof is now by induction to change points in the machine states.

By assumption, the proposition holds for t=0 (note that $R_i'(0)=R_i(0)$, for all $i\in M$, see (3.8))

Assume the assertions of the proposition hold until the change point in the state of machine i, t, say. At time t, the projected change point t_2 in the state of machine i is the same in N and N', since in both networks the machine states are equal and, in case of slow down of machine i, also the amount and duration of slow down of machine i are equal. Hence $t_2'=t_2$. Let t_1 be the next change point, either in a machine state or in a buffer content, in *N*. There are now two possibilities.

1. t_1 is a change point in the state of a machine j, say. Since all change points, machine states and machine speeds are the same in *N* and *N'*, prior to t, by the induction hypothesis, it follows by (3.6) that also all projected change points in machine states are the same in *N* and *N'*. Hence, $t_1'=t_1$ and the machine states are identical in *N* and *N'* between t and t_1. Since all machine speeds are equal in *N* and *N'* at time t and since they do not change until t_1, it follows that $X_j'(t_1)=X_j(t_1)$, for all $j\neq\ell$, and $X_\ell'(t_1)=K_\ell-X_\ell(t_1)$. By (3.8) it now follows that also $w_i'(t_1)=w_i(t_1)$, for all $i\in M$.
2. t_1 is a change point in the state of buffer j. Similarly as under 1., it follows by the induction hypothesis that $t_1'=t_1$ and $\{a_i'(t_1)\}=\{a_i(t_1)\}$ for all $i\in M$. Assume, that j becomes empty at t_1 in *N*. Since the machine speeds are equal in *N* and *N'* at time t and since the machine speeds do not change between t and t_1, it follows that j also becomes empty at $t_1'=t_1$ in *N'*, if $j\neq\ell$. If $j=\ell$, then j becomes full at t_1 in *N'*. Hence $X_r'(t_1)=X_r(t_1)$, for all $r\neq\ell$, $X_\ell'(t_1)=K_\ell-X_\ell(t_1)$. By (3.8) it now follows that $w_i'(t_1)=w_i(t_1)$, for all $i\in M$. Hence, the assertions of the proposition hold at time t_1. At time t_1, new projected change points in the states of the slowed down machines are generated in *N* and *N'*. By (3.6) it follows that these change points coincide in *N* and *N'*. We can now take the next change point t_3, and show in a similar way that the assertions of the proposition hold at t_3. By assumption, the change points do not condense. We can proceed in this way, until we arrive at a change point in a machine state.

In both cases, it follows that the assertions of the proposition hold at the next following change point in a machine state. QED

Proposition 3.1 implies that reversing the flow direction in a single buffer in an a/d network yields a sample–path–equivalent network. Hence a/d networks are sample–path–reversible. As a corollary, we can conclude that if a unique equilibrium buffer contents distribution exists for an a/d network *N* then, for every network *N'* that arises from *N* by reversing the flow in buffer ℓ, we have

$$P[X_j' \leq x] = P[X_j \leq x], \text{ for } j\neq\ell, \text{ and } P[X_\ell' \leq x] = P[X_\ell \geq K_\ell-x].$$

Note that, even when an equilibrium distribution exists, it need not be unique. It may depend on the initial state of the system.

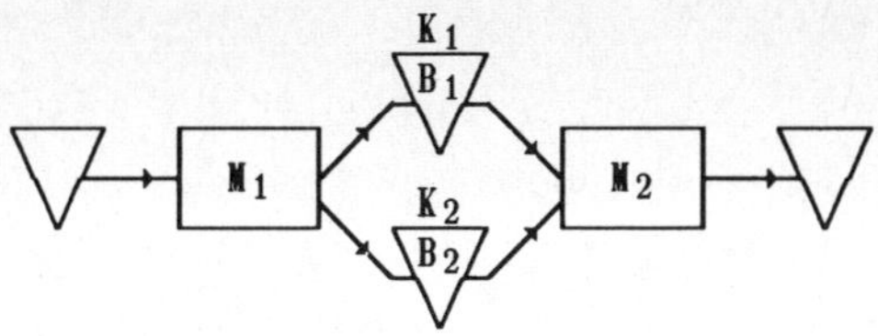

Figure 3.4.3. 2–machine a/d network

In the network in Fig. 3.4.3, machine M_1 performs disassembly and machine M_2 performs assembly. The difference in buffer contents remains constant. The initial state $X_1(0) = K_1$, $X_2(0) = 0$ in this figure leads to

$$\lim_{t\to\infty} X_1(t) = K_1, \lim_{t\to\infty} X_2(t) = 0.$$

the initial state $X_1(0) = 0$, $X_2(0) = K_2$ leads to

$$\lim_{t\to\infty} X_1(t) = 0, \lim_{t\to\infty} X_2(t) = K_2.$$

It is clear that networks as in Fig. 3.4.3 are only more or less realistic if $K_1 = K_2 =: K$ and if we start with $X_1(0)=X_2(0)$. However, whenever this is the case the system of Fig. 3.4.3 is equivalent to an ordinary flow line containing only one intermediate buffer of capacity K.
By means of proposition 3.1, directed cycles in a/d networks can be replaced by "bypasses", in order to have the flow moving in one direction only (from left to right). If the network then contains structures as in Fig. 3.4.3 (that is, without intermediate machines in the bypass), these can be eliminated, too, under the conditions mentioned before. For example, the cycle at buffers B_7 and B_{13} in Fig. 3.4.1 can be replaced by a "bypass", by changing M_7 into a disassembly machine, reversing the flow through B_{13}, and adding a source buffer supplying M_7.

3.4.2 Buffer–sharing networks

In this section, continuous–flow networks of linked buffers and machines are considered, where several machines obtain products from the same buffer and supply the same buffer. Each machine has only one upstream and one downstream buffer. Such a network is called *buffer–sharing.* A complex example of a buffer–sharing network is sketched in Fig. 3.4.4. Since several machines may share a common finite–capacity buffer, we have to do with a global restriction, holding for all machines, rather than a local restriction, holding for each machine separately. In general, the installation of such a common buffer, preceding several machines, improves the performance of the system, compared with the situation in which each machine has its own buffer, with the same total buffer capacity. If products are allocated to many buffers each supplying a different machine, then either of these machines stops if its preceding buffer is empty, whereas in other buffers there still may be products available.

If several machines obtain parts from the same buffer and that buffer is empty, then the total buffer output rate has to be reduced. It has to be decided what machines are allowed to operate and at what rates. A similar situation arises when several machines supply a single buffer and the buffer is full.

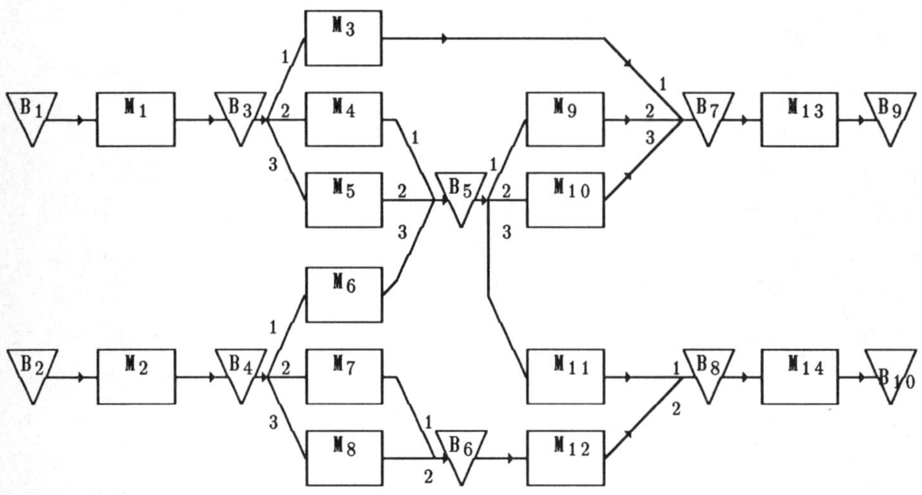

Figure 3.4.4. Buffer–sharing network

Assume buffer $b \in B$ is full and let it be preceded by machines $n_1,..n_r$ and succeeded by machines $m_1,..,m_s$. Let the potential speed of machine n_i be denoted by $v(n_i)$. If the total potential speed of all machines $n_1,..,n_r$ (σ_1, say, where σ_1 equals $\sum_{i=1}^{r} v(n_i)$) is greater than the total speed of machines $m_1,..,m_s$ (σ_2, say) at that moment, then machines $n_1,..,n_r$ will have to slow down until their total speed equals the speed of machines $m_1,..,m_s$. The reduction in the net speeds of the machines $n_1,..,n_r$ due to this is determined by an *allocation rule*.

Note that, in subsection 2.4.2, such speed allocation rules were not discussed for the parallel–machines case. This is because, in the case of a single buffer and two groups of parallel machines, the throughput and buffer contents distribution, are insensitive to the particular allocation rule (at least in the case of time–dependent failures). Only if the network contains more than one buffer, does the allocation rule become important.

An example of such an allocation rule is the following

R_1 Define priorities for each arc connecting a machine n_i with downstream buffer b as follows. The priority of the arc (n_i,b) is greater than the priority of (n_j,b) if $n_i<n_j$. If b is full and $\sigma_2<\sigma_1$ (terminology as above), then allocate the total remaining material flow among the machines $n_1,..,n_r$ in order of decreasing priorities. The actual speed of machine n_i, $w(n_i)$ now becomes
$\min \{v(n_i), \sigma_2 - \sum_{n_j<n_i} w(n_j)\}$.
If $\sigma_2>\sigma_1$ and the buffer is empty, then the machine speeds are determined similarly.

Note that this allocation rule is locally defined. In buffer–sharing networks, the allocation rule must be applied at each full or empty buffer. This leads, for all machines, to a number of equations that their speed have to satisfy. As an example, consider the network sketched in Fig. 3.4.5, with the indicated potential speeds and buffer contents. The allocation rule R_1 results in the following speeds:
$w(M_1) = \min\{4,2\} = 2$, $w(M_2) = \min\{1,w(M_1),2\} = 1$,
$w(M_3) = \min\{3,2-w(M_1),2-w(M_2)\} = 0$.

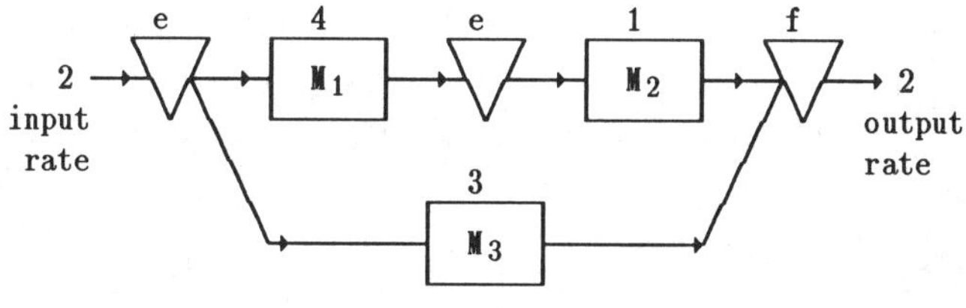

Figure 3.4.5. Four–machine network. Above machine M_i, $v(M_i)$ is indicated. For a buffer, e denotes that the buffer is empty, f denotes that the buffer is full.

The following is another allocation rule:

R_2 Allocate the reduction in potential speeds of the machines $n_1,..,n_r$, in the case that the downstream buffer b is full, equally over these machines. The speed of machine n_i then becomes

$$w(n_i) = \frac{v(n_i)}{\sum_{j=1}^{r} v(n_j)} \cdot \sigma_2.$$

For empty buffers the allocation rule works similarly.

For the network in Fig. 3.4.5 this allocation rule results in the following speeds

$$w(M_1) = 8/7, \; w(M_2) = \min\{w(M_1),1/2\} = 1/2, \; w(M_3) = \min\{6/7,6/4\} = 6/7.$$

Note that rule R_2 performs better than rule R_1, since $w_1(M_1)+w_1(M_3) = w_2(M_1)+w_2(M_3)$ and $w_1(M_2)+w_1(M_3) < w_2(M_2)+ w_2(M_3)$, where $w_i(M)$ is the speed of machine M under rule R_i. The best rule for this particular situation would be $w(M)=1$, for all machines M.

Many other allocation rules are possible. However, it may be possible that they lead to inconsistent systems of equations for the speeds $\{w_i(t);\; i\in M,\; t\geq 0\}$, or to systems with multiple solutions. In appendix D it is shown that, for networks with allocation rule R_1, with some straightforward extra restrictions, the system of equations for the machine speeds $\{w_i(t)\}$ at an arbitrary point in time is solvable and has a unique solution. An efficient algorithm based on dynamic programming for obtaining the machine speeds is discussed.

Now a proposition is presented for mixtures of buffer–sharing and a/d networks, analogous to proposition 3.1. A machine that has multiple upstream or downstream buffers in such a network performs assembly or disassembly, respectively. For each buffer with multiple upstream and downstream machines an allocation rule is needed. We do not give allocation rules for such networks, since only plain a/d and plain buffer–sharing networks are studied in the rest of this text. However, it is not difficult to generalize rules R_1 and R_2 for such combined networks.

Let N be a mixture of a buffer–sharing and an a/d network and let N' be the network arising from N by reversing the flow of buffer $\ell\in B$. If this results in a machine without an upstream buffer, then this machine becomes a disassembly machine and a source buffer is added for supplying this machine. If the flow reversal results in a machine without a downstream buffer, then this machine becomes an assembly machine supplying a sink buffer. Hence, the resulting network is again a mixture of a buffer–sharing and an a/d network. Starting from a plain buffer–sharing network, flow reversal only results in another buffer–sharing network, if the flows in *all* buffers are reversed.

The speed allocation rule in mixtures of buffer–sharing and a/d networks has to satisfy certain conditions.

Let *N* and *N'* be networks as above. An allocation rule R in a network *N*, which is a mixture of a buffer–sharing and a/d network, is said to possess the *local reversibility property*, if,

1. for all realizations of machine states $\{(t,a_i(t)) \mid t \in \mathbb{R}_+,\ i \in M\}$ and for potential machine speeds $\{(t,v_i(t)) \mid t \in \mathbb{R}_+,\ i \in M\}$, the used allocation rule R results in a *unique* set of real machine speeds $\{w_i(t) \mid t \in \mathbb{R}_+,\ i \in M\}$
2. If at an arbitrary time t, the realization $\{a_i'(t)\} = \{a_i(t)\}$, for all i and $X_j'(t) = X_j(t)$, $j \neq \ell$, $X_\ell'(t) = K_\ell - X_\ell(t)$, then allocation rule R determines the real machine speeds so that $w_i'(t) = w_i(t)$, for all i.

Note that for the case that *N* and *N'* are plain buffer–sharing networks, both R_1 and R_2 possess the local reversibility property.

For mixture networks *N* and *N'* as defined above, we assume that

- the assumptions for the machine states are the same as for a/d networks.
- The used allocation rule R possesses the local reversibility property.

Proposition 3.2 (sample–path–reversibility of mixed buffer–sharing a/d networks)
Let $\{(t,a_1(t),..,a_n(t));\ t \geq 0\}$ be a stochastic realization of the machine states in *N*, $\{(t,X_1(t),..,X_m(t));\ t \geq 0\}$ the corresponding realization of the buffer contents and $\{(t,w_1(t),..,w_n(t));\ t \geq 0\}$ the corresponding realization of the machine speeds in *N*.

Define $\{X_j'(0)\} = \{X_j(0)\}$, $j \in B$, $j \neq \ell$, $\{X_\ell'(0)\} = \{K_\ell - X_\ell(0)\}$.
$\{a_i'(0)\} = \{a_i(0)\}$, $i \in M$

Then,

$\{(t,a_1(t),..,a_n(t));\ t \geq 0\}$ is a stochastic realization of the machine states in *N'*, $\{(t,X_1(t),..,K_\ell - X_\ell(t),..,X_m(t));\ t \geq 0\}$ the corresponding realization of the buffer contents and $\{(t,w_1(t),..,w_n(t));\ t \geq 0\}$ the corresponding realization of the machine speeds *N'*.

Sketch of the proof. The proof is along the same line as that of proposition 3.1. Suppose the assertions of the proposition hold until the change point in a machine state t. Let t_1 be the next change point, either in machine states or in buffer contents, in *N*. As in the proof of proposition 3.1, there are two possibilities, of which only one is treated:

Assume that t_1 is a change point in the state of buffer j. Assume, without loss of generality, that j becomes empty at t_1. Similarly as in the proof of proposition 3.1, it follows by the induction hypothesis that $t_1'=t_1$ and $\{a_i'(t_1)\}=\{a_i(t_1)\}$ for all $i \in M$. Since the real machine speeds in *N* and *N'* are equal at t, and since they do not change until t_1, it follows that j also becomes empty at t_1 in *N'*, if $j \neq \ell$ and j becomes full at t_1 in *N'*, if $j=\ell$. Hence $X_j'(t_1) = X_j(t_1)$, for all $j \neq \ell$, $X_\ell'(t_1) = K_\ell - X_\ell(t_1)$. By the local reversibility property of the allocation rule R, it follows that $w_i'(t_1)=w_i(t_1)$, for all $i \in M$. Hence, the assertions of the proposition hold at time t_1. We can now take the next change point t_2, and show in a similar way that the assertions of the proposition hold at t_2. By assumption, the change points do not condense. We can proceed in this way, until we arrive at a change point in a machine state. QED

In Corten and De Koster (1987) discrete product a/d and buffer–sharing networks are also considered. If the blocking rule is chosen properly, it appears to be possible to prove analogues of propositions 3.1 and 3.2 for such discrete–product networks.

CHAPTER 4

ANALYSIS OF LOCALLY CONTROLLED SYSTEMS

In the previous chapters, several types of single–product multistage production networks have been introduced. Tools are presented in Chapters 2 and 3 for the analysis of two and three–stage systems. Larger systems are not tractable analytically, at least not exact. Therefore, other tools are needed for the evaluation of larger networks. A possible approach is by simulation and heuristic analytic methods. As indicated in Chapter 3, simulation of large networks is time–consuming. For that reason, heuristic methods are developed in this chapter for evaluating production networks with several buffers. These tools have the advantage over simulation that they provide quick insight into relationships between system performance and buffer sizes. See the discussions in sections 1.4 and 3.1. Such tools can then be used in the design of production systems and production–control rules. We are mainly interested in aggregate performance criteria of the system, such as throughput, average work–in–process and average throughput time.

In general production networks, we may have to deal with many stages, many products and all kinds of possible control rules (integral or local). In this chapter and the following one, first single–product, multistage networks with different layouts are considered. Furthermore, only local control rules are considered here. The nature of the local control is simple: it is caused by finite buffer capacities. Only in section 4.5 more complicated $<m,M>$ switching polices are considered.

In the next chapter some results will be extended to single–product, integrally controlled production lines. In Chapter 6, applications of the approximation method to more general multiproduct systems is discussed.

To approximate the performance of a production system, two–stage subsystems, consisting of two machines and a finite intermediate buffer, are repeatedly replaced by a single aggregate production stage. This leads to smaller production systems which eventually can be evaluated analytically.

The aggregation step used in the approximation method is explained in section 4.1. Aggregation of two production stages and an intermediate finite buffer has been applied by many authors. Examples can be found in Buzacott (1967), Gershwin (1987), Terracol and David (1987a, 1987b), Ancelin and Semery (1987), Hontelez (1985). Nearly all heuristic analytic literature mentioned in subsections 3.2.1 and 3.2.2 deals in fact with aggregation methods. In section 4.2 the approximation method for flow lines is explained. Numerical results are given in section 4.3. A comparison is given with another approximation method for continuous–product flow lines, developed by Glassey and Hong (1986b). The approximation method and the obtained results are based on De Koster (1987b, 1988b). In section 4.5, the aggregation method is tested on flow lines where the last buffer is controlled with a switching–point policy, as described in section 2.5.2.

It is a straightforward matter to extend the approximation method to locally controlled production networks with more complicated layouts. Assembly–disassembly networks are investigated in subsection 4.4.1. Results in this subsection are based on De Koster (1987c). In subsection 4.4.2, the approximation method is applied to buffer–sharing networks. These results are based on De Koster (1988a). In section 4.6, a final evaluation of the material of this chapter is given.

4.1 AGGREGATION OF TWO–STAGE LINES

The approximation method for continuous–flow production systems is based on the substitution of two–stage subsystems by one stage. This is called *aggregation*. This aggregation will be worked out first for simple up–down machines. Then it will be extended to more general machines like those described in section 2.4.

a. Two–state unreliable machines

Assume that the two–stage line is as sketched in Fig. 4.1.1.

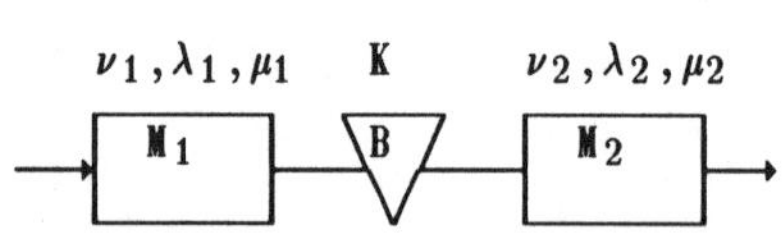

Figure 4.1.1 Continuous–flow two–stage line

The line in Fig. 4.1.1 is of the type described in section 2.3, with a finite buffer of capacity K. The output behavior (and the input behavior, too) of this line is of a type similar to that of an individual machine, namely with a pattern of up and down periods. These periods are however more complex than those of a single machine. A period of no output may be caused by a failure of M_2, but also by a failure of M_1, if the buffer is empty at the same time. If there is a positive output, the line may run at rate ν_2, but also at rate ν_1, if $\nu_1 < \nu_2$ and the buffer is empty. It is possible to approximate the output behavior in various ways. A simple way is to approximate the down and up times by exponential distributions and the speed by the average output speed, provided there is output. In this way the aggregate stage is of the same type as the original line, with speed ν_{out}, failure rate λ_{out} and repair rate μ_{out}. The parameters ν_{out}, λ_{out} and μ_{out} can be determined by using quantities defined in Table 2.3.1. These quantities are calculated as indicated in Table 2.3.2. The parameter λ_{out}, for example, is determined as the quotient of the average number of switches per cycle from a positive line output speed to zero output (O_o) and the average total time per cycle that there is line output ($S_{\nu_1}+S_{\nu_2}$). For the exact definition of these parameters, in terms of quantities defined in Table 2.3.1, see Table 4.1.1.

	machine speed	failure rate	repair rate
output	$\frac{\nu_1 S_{\nu_1}+\nu_2 S_{\nu_2}}{S_{\nu_1}+S_{\nu_2}}$	$\frac{O_o}{S_{\nu_1}+S_{\nu_2}}$	$\frac{O_o}{S_0}$
input	$\frac{\nu_1 I_{\nu_1}+\nu_2 I_{\nu_2}}{I_{\nu_1}+I_{\nu_2}}$	$\frac{O_i}{I_{\nu_1}+I_{\nu_2}}$	$\frac{O_i}{I_0}$

Table 4.1.1 Determination of (ν,λ,μ) of the aggregate stage

Note that, for the choice ν_{out}, λ_{out} and μ_{out} of the parameters of the aggregate stage as in Table 4.1.1, the throughputs of the aggregate stage and the original two–stage line are equal. (This follows from the fact that the throughput of the two–stage line equals

$$v(K) = \frac{\nu_1 S_{\nu_1}+\nu_2 S_{\nu_2}}{S_0+S_{\nu_1}+S_{\nu_2}} = \frac{\nu_{out}\mu_{out}}{\lambda_{out}+\mu_{out}}.$$

A similar reasoning can be followed for the input behavior of the two–stage line. Sometimes there is a positive input and sometimes the line has no input. This input behavior can also be approximated. This results in the parameters of the aggregate stage ν_{in}, λ_{in}, μ_{in}. The choice of the input parameters is also indicated in Table 4.1.1. Note that for a choice as indicated in Table 4.1.1, it holds that

$$v(K) = \frac{\nu_1 I_{\nu_1} + \nu_2 I_{\nu_2}}{I_0 + I_{\nu_1} + I_{\nu_2}} = \frac{\nu_{in}\mu_{in}}{\lambda_{in}+\mu_{in}}.$$

These input and output parameters of the aggregate stage need not be close to each other. The output parameters will be close to those of M_2, the input parameters will be close to those of M_1. In general, the output behavior is used if it is desired to approximate the behavior of the content of downstream buffers of the two–stage line. The input behavior is used to approximate the behavior of the content of upstream buffers.

b. General N–state machines

It is also possible to aggregate two–stage lines with multistate machines, as described in section 2.4, to one two–state machine. However, the states of this aggregate are not necessarily chosen to be up and down states. The states are determined by considering some speed level z, so that the output speed of the two–stage line is sometimes greater than z and sometimes less than or equal to z. This level z can be chosen arbitrarily, under the condition that the output rate has to be greater than z and less than or equal to z, for a fair amount of time. The system is said to be in state 1, if the output speed is greater than z, and in state 2, if the output speed is less than, or equal to z. The duration of speed > z and speed $\leq$ z intervals are approximated by exponential distributions. The parameters of the distributions can be calculated, using results of section 2.4. In Table 2.4.1 it is indicated how the cost rates $\alpha_i(x)$ (see section 2.4) have to be chosen, needed for the calculation of the average number of switches per cycle, from a speed > z to a speed $\leq$ z (O_0^z) and the total time per cycle of output speed > z ($\sum_{\sigma>z} S_\sigma$). The quotient of these two quantities determines the rate of the exponential distribution of state 1 (output speed > z) of the aggregate stage.

The input behavior of the two–stage line can be approximated in the same way. In Table 4.1.2 all parameters of the aggregate stage are defined in terms of quantities defined in Table 2.3.1 and section 2.4. By a choice of the parameters as indicated in this table, again the production rate of the two–stage line equals the the average speed of the aggregate machine, or, in formula

$$v(K) = \sum_\sigma \sigma S_\sigma / \sum_\sigma S_\sigma = \frac{\nu_{out,1}\lambda_{out,2}+\nu_{out,2}\lambda_{out,1}}{\lambda_{out,1}+\lambda_{out,2}}$$

$$= \sum_\sigma \sigma I_\sigma / \sum_\sigma I_\sigma = \frac{\nu_{in,1}\lambda_{in,2}+\nu_{in,2}\lambda_{in,1}}{\lambda_{in,1}+\lambda_{in,2}}$$

	machine speed 1	machine speed 2	life rate 1	life rate 2
output	$\dfrac{\sum_{\{\sigma>z\}} \sigma S_\sigma}{\sum_{\{\sigma>z\}} S_\sigma}$	$\dfrac{\sum_{\{\sigma\leq z\}} \sigma S_\sigma}{\sum_{\{\sigma\leq z\}} S_\sigma}$	$\dfrac{0_o^z}{\sum_{\{\sigma>z\}} S_\sigma}$	$\dfrac{0_o^z}{\sum_{\{\sigma\leq z\}} S_\sigma}$
input	$\dfrac{\sum_{\{\sigma>z\}} \sigma I_\sigma}{\sum_{\{\sigma>z\}} I_\sigma}$	$\dfrac{\sum_{\{\sigma\leq z\}} \sigma I_\sigma}{\sum_{\{\sigma\leq z\}} I_\sigma}$	$\dfrac{0_i^z}{\sum_{\{\sigma>z\}} I_\sigma}$	$\dfrac{0_i^z}{\sum_{\{\sigma\leq z\}} I_\sigma}$

Table 4.1.2 Parameters of the aggregate stage, in case of stages of the type of section 2.4

Note that there are no explicit breakdowns if z>0. Furthermore, the quantities expressed in Table 4.1.1 are special cases from those of Table 4.1.2 (z=0).

Other ways of approximating two–stage lines as in Fig. 4.1.1 are described in De Koster (1986b). However, in the present case only the approximation by a two–state stage, with parameters as expressed in Table 4.1.1 (or for more complex stages, as in Table 4.1.2) is investigated. It will be shown that this approximation is quite satisfactory.

4.2. APPROXIMATION OF FLOW LINES

In this section it will be shown how the behavior of flow lines, as sketched in Fig. 4.2.1, is approximated, using aggregation as introduced in the previous section.

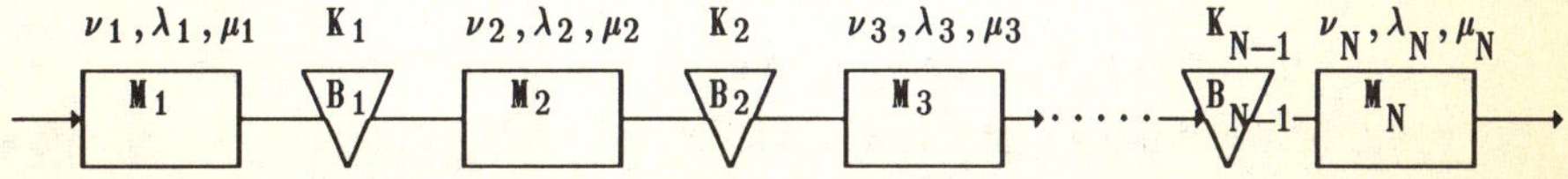

Figure 4.2.1 N–stage flow line. λ_i is the failure rate, μ_i the repair rate, ν_i the speed of M_i. K_i is the capacity of B_i.

The line sketched in Fig. 4.2.1 is locally controlled by finite buffers. Each machine reacts to the immediately preceding and following machine (through empty or full buffers, respectively). A description of the operation of the line is given in subsection 3.3.1.

4.2.1 Approximation method 1

A straightforward method (from now on denoted by approximation method 1, or AP1) for approximating the throughput of the line and simultaneously the behavior of the contents of B_{N-1}, is by using the scheme in Fig. 4.2.2.

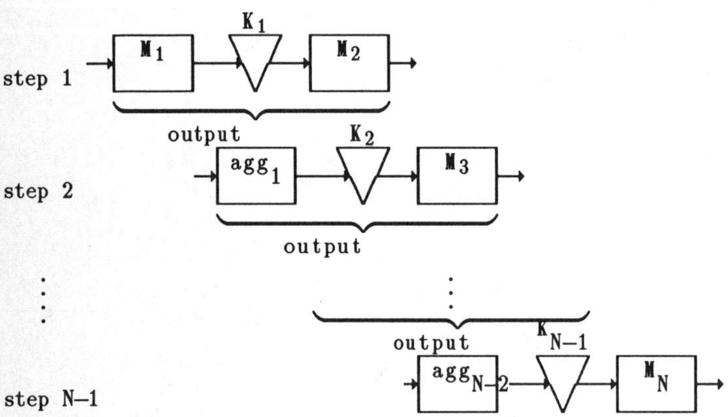

Figure 4.2.2 Approximation from left to right of the line in Fig. 4.2.1

The aggregate machine in step i, denoted by agg_i, is defined by the parameters determined from the output behavior of the two-stage line obtained in step i–1. The parameters of the aggregate machine are determined as described in the previous section. The first machine of the two-stage line is an aggregate by itself. In step N–1, the obtained throughput and mean buffer contents of the two-stage line approximate the throughput and mean buffer contents of B_{N-1} of the original line. To obtain estimates for the behavior of B_1, the scheme has to be applied from right to left. Instead of output behavior, input behavior is calculated at each step. To obtain estimates for the behavior of other buffers, combination of approximation from left to right and approximation from right to left has to be applied. For example, if the mean buffer contents of B_i $(2 \leq i \leq N-2)$ is desired, approximation from left to right can be applied on the line M_1–B_1–M_2–$\cdots$–M_i, resulting in agg_{i-1}. Approximation from right to left of $M_{i+1}-B_{i+1}-\cdots-B_{N-1}-M_N$ results in agg_{i+1}. Evaluation of $agg_{i-1}-B_i-agg_{i+1}$ gives an estimate for the behavior of B_i.

Numeric evaluations for some small three–stage production systems, are given in Table 4.2.2. The parameter choices of the lines, given in Table 4.2.1, are motivated as follows:

1. Balance vs unbalance. A line is called *balanced* if all machine efficiencies are equal. All lines, except case 5, are unbalanced. Case 2 is nearly balanced. Case 4 is highly unbalanced.
2. The repair rates are greater than the failure rates. M_2 in case 1 is perfect. Buffer capacities vary from small (case 1) to large (case 5).
3. Location of bottleneck and fastest machine. In cases 1, 2 and 3, the bottleneck is in the middle, the position of the fastest machine varies. In case 4, the fastest machine is in the middle, the bottleneck is the first machine.

case	N	ν	λ	μ	$\mathbf{K}$	eff
1.	3	(1.2,1,1.2)	(.01,0,.005)	(.09,1,.09)	(5,5)	(1.08,1,1.14)
2.	3	(1.2,1,1.5)	(.02,.01,.025)	(.08,.09,.04)	(10,10)	(.96,.9,.92,
3.	3	(1,0.9,1)	(.01,.01,.01)	(.09,.07,.09)	(45.49,20)	(.9,.7875,.9)
4.	3	(0.9,4,1)	(.01,.02,.01)	(.09,.08,.09)	(10,10)	(.81,3.2,.9)
5.	3	(1,1.1,1.2)	(.01,.02,.03)	(.09,.09,.09)	(50,50)	(.9,.9,.9)

Table 4.2.1. Parameters of five three–machine lines

In Table 4.2.2, X_i stands for the buffer content of the i–th buffer, v(**K**) stands for throughput. In that table, approximation results are compared with simulation. E stands for the expectation operator, in case of the numerical evaluations; it is used to denote averages in the case of simulation results. Unless otherwise stated, all simulation results in this chapter are obtained by 10 runs of 500,000 units of time per run on a Burroughs B7900 computer and all lifetimes are assumed to be time–dependent. The simulation program is event–driven, which means that machine failures and repairs are generated, after which the buffer behavior is calculated. Such an approach has also been described by D'Angelo et al. (1988). Relative errors are measured in the usual way (note that in section 2.7 relative errors are measured differently). The approximate total work–in–process in the system, denoted by EL, is obtained by summing EX_1, obtained by approximation from right to left and EX_2, obtained by approximation from left to right.

case method	v(K)	ϵ	EX_1	ϵ	EX_{N-1}	ϵ	EL	ϵ
1. sim.	.9019(5)		4.205(3)		0.403(2)		4.607(6)	
AP1 (l)	.9017	0.0			0.400	0.5	4.614	−0.2
AP1 (r)	.9018	0.0	4.214	−0.2				
2. sim.	.6036(4)		7.709(11)		3.912(7)		11.62(2)	
AP1 (l)	.5939	1.6			3.866	1.2	11.65	−0.3
AP1 (r)	.5958	1.3	7.782	−0.9				
3. sim.	.7650(7)		38.11(8)		4.380(22)		42.49(9)	
AP1 (l)	.7648	0.0			4.423	−1.0	42.64	−0.4
AP1 (r)	.7652	−0.0	38.22	−0.3				
4. sim.	.7162(2)		2.469(3)		4.406(3)		6.876(5)	
AP1 (l)	.7040	1.7			2.940	33.3	6.969	−1.4
AP1 (r)	.6891	3.8	4.029	−63.2				
5. sim.	.8288(6)		30.60(14)		22.63(11)		53.23(13)	
AP1 (l)	.8204	1.0			21.30	5.9	53.01	0.4
AP1 (r)	.8242	0.6	31.71	−3.6				

Table 4.2.2 Numeric results for the lines in Table 4.2.1. ϵ is the relative error compared with simulation (in %). AP1 (l) stands for approximation from left to right, AP1 (r) stands for approximation from right to left.

For the simulation results confidence intervals are indicated. The notation .9019(5) means that the 95% confidence interval (using the student distribution with 9 degrees of freedom) is given by (.9014,.9024).

The approximation results for the first three cases and case 5 are good, but the mean buffer–content approximations for the fourth case are not good (although the total system–content approximation is good). Also, the throughput approximation shows a higher relative error than in the other cases. This is due to the fact that M_2 is much faster on average than M_1. M_2 is so fast that, as long as M_2 is up, M_1 has an available buffer of capacity K_1+K_2. It occurs relatively often that both buffers are full simultaneously, and hence M_1 is frequently blocked not by M_2, but by M_3. Similarly, it frequently occurs that M_3 is slowed down by M_1, since both B_1 and B_2 are empty. Compare this situation with the extreme situation that M_2 is perfect and $\nu_2 \geq \max\{\nu_1,\nu_3\}$ (see section 3.3.2.2). This situation is exactly equivalent to M_1–B–M_3, with B a buffer of capacity K_1+K_2. The approximation scheme, applied from left to right, yields M_1–B_2–M_3 in this case (since aggregation of M_1–B_1–M_2 yields M_1), the throughput of which may severely underestimate the real throughput (because $K_2<K_1+K_2$). Also, for most other cases in Table 4.2.2, the approximation method underestimates the throughput (except 3).

Therefore, in the next subsection the approximation scheme is adapted for three–stage lines with fast second machines. An adapted scheme for N–stage lines is given in subsection 4.2.3.

4.2.2. Approximation method 2 for three–stage lines

In this section, the approximation method AP1 is adapted for three–stage lines with fast second machines. The adapted method is denoted by AP2. Assume M_2 is faster on average than M_1, or $eff_2 > eff_1$. In this case the scheme in Fig. 4.2.2 underestimates the throughput (although throughput estimates are still not bad). The mean content of B_2 is underestimated, the mean content of B_1 is over–estimated. Depending on the difference in machine efficiencies of M_1 and M_2, the buffer capacity in the second approximation step has to be closer to K_1+K_2 than to K_2, to obtain a better estimate of the throughput.

Therefore we propose the following approach in approximating from left to right in case the first machine is frequently blocked by the last one. This will usually occur when the second machine is faster on average than the first one, hence if $eff_2>eff_1$, or if $eff_2=eff_1$ and M_2 is more stable than M_1. The approximation consists of two parts. The first part is the estimation of the throughput, the second part the approximation of the behavior of the contents of B_2. The first part is sketched in Fig. 4.2.3.

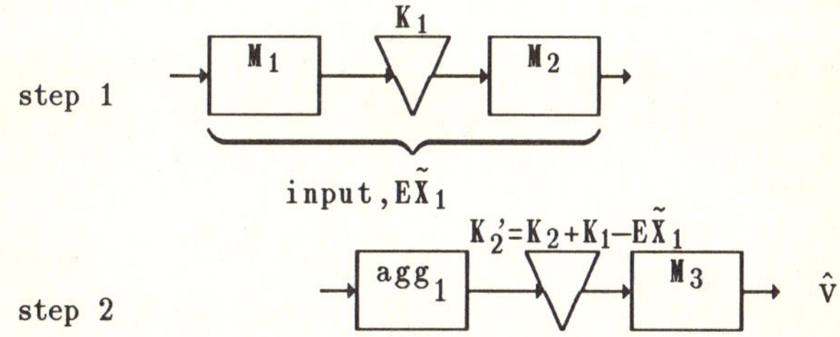

Figure 4.2.3 Approximation from left to right in the case that $eff_2 > eff_1$

In step 1 in Fig. 4.2.3 the average buffer contents, $E\tilde{X}_1$, as well as the parameters of agg_1 are calculated. Instead of *output* behavior, the *input* behavior is used for

the determination of the parameters of the aggregate (see section 4.1). The reason for this is, that the total line content downstream of the input is to be considered. In step 2, $K_2+K_1-E\tilde{X}_1$ is taken as buffer capacity, since the input of the line in step 2 has still $K_1-E\tilde{X}_1$ buffer capacity available in agg_1 and additionally K_2 in B_2. This gives us an estimate for the throughput, $\hat{v}$, say.
That this estimate for the throughput is a good one in the case that $eff_2>eff_1$, can be understood from the following special case (see also subsection 3.3.2).

If M_2 is perfect and $\nu_2 \geq \max\{\nu_1,\nu_3\}$, then $E\tilde{X}_1=0$ and hence $K_2'=K_1+K_2$. Furthermore, agg_1 equals M_1, hence the approximation is exact in this case.

For the cases where method AP1 performs well, method AP2 also performs reasonable. This can be seen from the following example.
Assume M_3 is perfect and $\nu_3 \geq \nu_2$. This line is equivalent to the line M_1–B_1–M_2. Hence the approximation of Fig. 4.2.3 again yields the exact throughput. Note also that the scheme in Fig. 4.2.2 is exact in this case.

The second part of the approximation consists of the approximation of the buffer content of B_2, by increasing the capacity of B_1 followed by application of the scheme in Fig. 4.2.2. Note that the approximation by the scheme in Fig. 4.2.3 cannot be used for an approximation of the content of B_2, since the buffer capacity in the second approximation step does not correspond to the buffer capacity of B_2 in the original line. The constant C that has to be added to K_1 has to be chosen such, that the obtained throughput estimate approximates $\hat{v}$ (see the scheme in Fig. 4.2.2). This constant C can be found iteratively. In such an iteration, a constant C_1 is added to K_1, then the scheme in Fig. 4.2.2 is applied, and this yields a throughput estimate. This throughput estimate is compared with $\hat{v}$. If it is smaller than $\hat{v}$, then C_1 is increased, if it is greater, then C_1 is decreased. This process is repeated until sufficient accuracy is reached. This yields an estimate for the mean content of B_2. In some cases it may be necessary to take C equal to ∞. This is the case, when all values of C_1 yield underestimates of $\hat{v}$.

In approximating from right to left a similar procedure is followed if $eff_2 > eff_3$. In Table 4.2.3 the performance of AP2 is given for the three–stage cases 4 and 5. The results in this table can be compared with results in Table 4.2.2. The indicated throughput estimate for method AP2, v(K), is obtained by adding such a constant C to the capacity of B_1, followed by approximation with the scheme in

Fig. 4.2.2, as indicated above. In case 4, method AP2 is applied both in the approximation from left to right and from right to left. Although the machine efficiencies are equal for case 5, the variance in machine speeds increases from left to right, which implies that the machines become more stable considered from right to left. Therefore method AP2 is only applied in the approximation from right to left.

case method	$v(K)$	ϵ	EX_1	ϵ	EX_{N-1}	ϵ	EL	ϵ
4. sim.	.7162(2)		2.469(3)		4.406(3)		6.876(5)	
AP2 (l)	.7288	−1.8			4.192	4.9	7.132	−3.7
AP2 (r)	.7056	1.5	2.940	−19.1				
5. sim.	.8288(6)		30.60(14)		22.63(11)		53.23(13)	
AP1 (l)	.8204	1.0			21.30	5.9	52.29	1.8
AP2 (r)	.8289	−0.0	30.99	−1.3				

Table 4.2.3. Numerical results for method AP2, for cases 4 and 5

Note that the relative errors in the throughput, but especially in the mean buffer contents have become significantly smaller by using method AP2.

4.2.3 Method AP2 for N–stage lines

The approximation for three–stage lines can be extended to N–stage lines (N>3). The effect of throughput underestimation by method AP1, will even be stronger in N–stage lines than in three–stage lines. This is because, in a long line, in which the machines are about equally fast, a machine far away from the beginning or end of the line, will be much faster on average than the whole line preceding or succeeding it. Recall that the throughput underestimation by method AP1 for three–stage lines, has to do with the fact that the middle machine is faster than the first or the last machine machine. In approximating the throughput from left to right in an N–stage line, the approximation results will be best when most buffers are adapted (depending on the efficiency of the machines), as was done with method AP2 for three–stage lines. However, for lines with more than three machines we do not use an iterative method to approximate the behavior of the contents of B_{N-1}. It is more convenient to adapt all buffer capacities except that of B_{N-1} and leave the capacity of B_{N-1} unchanged. This will be shown to be not serious. In Fig. 4.2.4 a scheme for the approximation of B_{N-1} and the throughput is given, inspired by the scheme in Fig 4.2.3. This approximation method for lines with more than three machines will also be denoted by AP2.

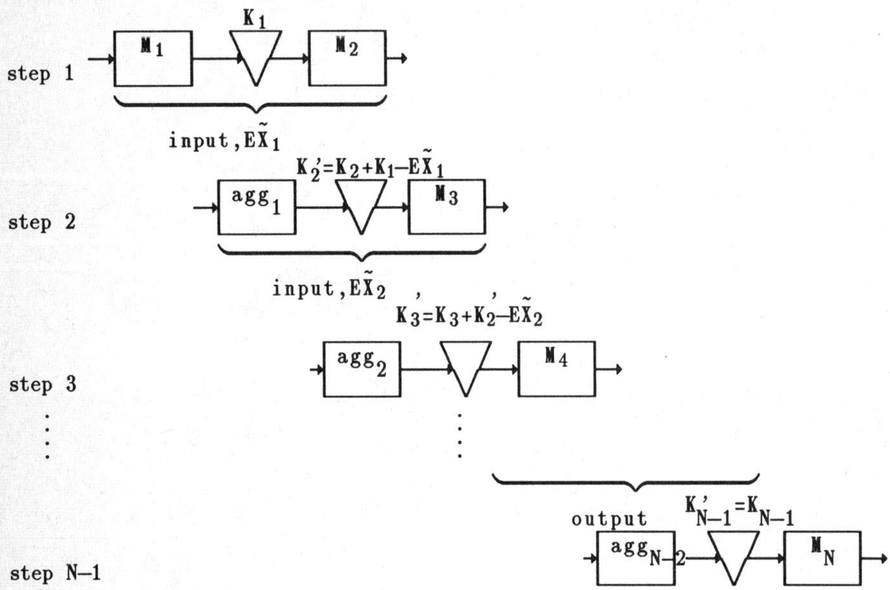

Figure 4.2.4 Approximation from left to right of the line in Fig. 4.2.1

In approximation step i in Fig. 4.2.4, the parameters of agg_i are calculated as well as $\tilde{EX}_i$, the mean buffer content of the two–stage line in this step. In this calculation the input behavior of the two–stage line is used, for the same reason as explained in section 4.2.2. In the last step the output behavior is used, as indicated in Fig. 4.2.4, since the aim is to approximate the behavior of the contents of B_{N-1}. The buffer capacity in the i–th step is K_i', where

$$K_1' = K_1$$

$$K_i' = K_i + K_{i-1}' - \tilde{EX}_{i-1}, \quad 2 \leq i \leq N-2$$

$$K_{N-1}' = K_{N-1}$$

The approximation from left to right, as sketched in Fig. 4.2.4 is for the approximation of the behavior of the contents of B_{N-1}. For the approximation of the contents of B_1 we have to approximate from right to left. In such an approximation a similar scheme as in Fig. 4.2.4 can be used. However, instead of input and output behavior, we then have to use output and input behavior, respectively. Furthermore, K_i' is defined as $K_i+E\tilde{X}_{i+1}$, except for the first and the last step. This scheme is sketched in Fig. 4.2.5.

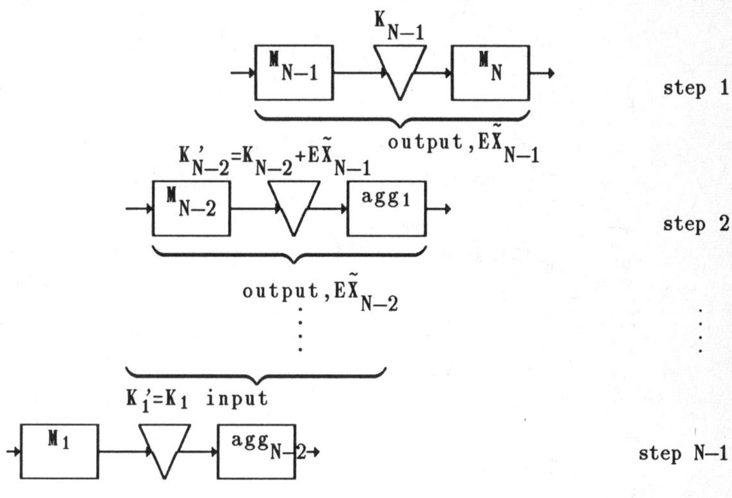

Figure 4.2.5. Approximation from right to left of the line in Fig. 4.2.1

If the behavior of the contents of another buffer is desired, say B_i, for $2\leq i\leq N-2$, then a combination of approximation from the left and from the right has to be used. For instance, let $3\leq i\leq N-3$, then approximation from left to right can be applied on the line $M_1-B_1-\cdots-B_{i-1}-M_i$ with K_{i-1}' defined as $K_{i-1}+K_{i-2}'-E\tilde{X}_{i-2}$, instead of K_{i-1}. Similarly, approximation from right to left can be applied on the line $M_{i+1}-B_{i+1}-\cdots-B_{N-1}-M_N$ with K_{i+1}' defined as $K_{i+1}+E\tilde{X}_{i+2}$. Combining the results of both approximation procedures yields an approximation of the behavior of the contents of B_i.

If the behavior of all buffer contents is desired, for example, to obtain an estimate for the mean system content, then it is efficient to approximate the line from left to right for B_{N-1}, then from right to left for B_1 and subsequently to approximate the contents of all intermediate buffers. In this way $2(N-1)+N-3= 3N-5$ two-stage lines need to be evaluated. From Table 2.3.6 it can be concluded that this takes less than (3N–5)/2 CPU seconds on a Burroughs B7900 computer, in a straightforward implementation.

Concluding, we can summarize method AP2 as follows (only the approximation from left to right is given, the approximation from right to left is similar).

1. N=3. For $eff_2 \leq eff_1$, the approximation scheme is given by Fig. 4.2.2. For $eff_2 > eff_1$, or $eff_2 = eff_1$ and M_2 more stable than M_1, the scheme is given by Fig. 4.2.3.
2. N>3. In this case, the scheme is given by Fig. 4.2.4.

4.3. MORE NUMERIC RESULTS AND COMPARISONS

In this section numeric results of the approximation methods are given. They are compared with simulation results on the one hand and with results of Glassey and Hong (1986b) on the other. The algorithm described in Glassey and Hong (1986b) is the continuous–flow version of the algorithm of Gershwin (1987). The algorithm of Gershwin is an approximation method for discrete product lines with constant cycle times, which are equal for all machines, and geometrically distributed life and repair times for the machines. It is based on a similar aggregation technique as method AP1, apart from the fact that all machine speeds are equal and the failure and repair rates of the aggregates are found iteratively. For the evaluation of all mean buffer contents of an N–stage line, Glassey and Hong (1986b) use a standard algorithm for the iterative solution of a system of 6N–5 nonlinear equations, which is similar to that of Gershwin (1987). Glassey and Hong tested their algorithm only for lines with state–dependent failures, here it is also tested for lines with time–dependent failures.

In Table 4.3.1, four lines are given, which are approximated by method AP2. The parameters in this table are not chosen to cover the whole scale of possible parameter settings, which would be a huge task (in an N–stage line there are 4N–1 parameters which may be varied). They form a small sample of more or less realistic parameter settings. The motivation of the parameter choices is the same as in section 4.2. Furthermore, the line length is varied. Approximation results are given in Table 4.3.2.

case	N	$\boldsymbol{\nu}$	$\boldsymbol{\lambda}$	$\boldsymbol{\mu}$	$\mathbf{K}$	eff
6	4	(1.2,1,1.5, 1)	(.02,.01,.025, .001)	(.08,.09,.04, .009)	(10,10,10)	(.96,.9,.92, .9)
7	4	(1,1.2,1.5, 1)	(.01,.02,.025, .001)	(.09,.08,.04, .009)	(10,10,10)	(.9,.96,.92, .9)
8	5	(1.2,1,1.5, 1,0.9)	(.02,.01,.03, .001,.005)	(.08,.09,.05, .009,.07)	(20,20,20, 20)	(.96,.9,.94, .9,.84)
9	5	(1,1.5,1.2, 1,0.9)	(.01,.03,.02, .001,.005)	(.09,.05,.08, .009,.07)	(20,20,20, 20)	(.9,.94,.96, .9,.84)

Table 4.3.1. Four multistage cases

case	method	v($\mathbf{K}$)	ϵ	EX_1	ϵ	EX_{N-1}	ϵ	EL	ϵ
6.	sim.	.5490(8)		7.911(9)		3.487(9)		16.02(2)	
	AP2 (l)	.5326	3.0			3.456	0.9	16.83	−5.0
	AP2 (r)	.5692	−3.7	7.876	0.4				
7.	sim.	.5511(9)		6.548(9)		4.012(10)		16.14(2)	
	AP2 (l)	.5564	−1.0			3.559	11.3	16.51	−2.4
	AP2 (r)	.5666	−2.8	7.248	−10.7				
8.	sim.	.6319(4)		15.46(2)		6.922(21)		41.31(6)	
	AP2 (l)	.6386	−1.1			5.685	17.9	42.36	−2.6
	AP2 (r)	.6613	−4.7	15.62	−1.0				
9.	sim.	.6284(7)		12.05(2)		6.850(26)		35.37(6)	
	AP2 (l)	.6462	−2.8			6.151	10.2	37.44	−5.9
	AP2 (r)	.6456	−2.7	12.68	−5.2				

Table 4.3.2. Approximation of the lines in Tables 4.3.1

In Table 4.3.2 confidence intervals are given as before. It is found that the approximation method performs reasonably for not too large systems. Note that relative errors in throughput are much smaller than relative errors in mean buffer contents. The first error is less than 5% for all systems, whereas the second error is 17.9% for the mean contents of B_{N-1} in one of the five-stage systems. More numerical examples can be found in De Koster (1987b) and De Koster (1988b). All these examples show that for not too unbalanced lines, relative errors in throughput are slight (<5% for lines up to 6 machines). Relative errors in average *total* buffer contents are greater (<10%, for lines up to 6 machines).

There is some literature on approximation methods for the above–presented model with continuous flow and unreliable machines. Glassey and Hong's paper has been mentioned already. Recently, Ancelin and Semery (1987) and Terracol and David (1987a, 1987b) presented an approximation method which resembles the method presented here. However, they only give results for the throughput.

In Table 4.3.3 five lines are listed to be compared with simulation and with the method of Glassey and Hong (denoted by G&H), for throughput and mean content of B_1 and B_{N-1}. Two lines have time–dependent failures and three lines have state–dependent failures, in the sense that machines that are blocked or starved cannot fail. Time–dependent failures and state–dependent failures are denoted by TD and SD, respectively. Results are given in Table 4.3.4.

case	N	ν	λ	μ	K	eff	SD/TD
10.	3	(1,1.1,1.2)	(.01,.02,.03)	(.09,.09,.09)	(50,50)	(.9,.9,.9)	SD
11.	3	(1.5,1,1.1)	(.05,.02,.03)	(.1,.08,.07)	(30,70)	(1,.8,.77)	SD
12.	4	(1.1,1,1, 0.9)	(.02,.01,.01, .01)	(.08,.07,.09, .05)	(40,50) 60)	(.88,.875, .9,.75)	SD
13.	3	(1,0.9,1)	(.01,.01,.01)	(.09,.07,.09)	(18.29,8)	(.9..7875,.9)	TD
14.	5	(0.9,1.05, 1,0.9,1)	(.01,.01,.01, .01,.01)	(.09,.07,.09, .07,.09)	(6.96,6.38, 10.29,8)	(.81,.919, .9,.7875,.9)	TD

Table 4.3.3. Five lines with time dependent and state–dependent failures

From the results in Table 4.3.4 it is found that algorithm AP2 and the method G&H approximate the throughput well, in general. Errors in mean buffer–content approximations are greater than those in throughput estimates. In many cases the algorithm of G&H often performs somewhat better than AP2 (especially for state–dependent failures). A major drawback however is the long and unpredictable run times needed. Table 4.3.4 lists the number of two–stage line evaluations for both algorithms. For an N–stage line (N>3), algorithm AP2 needs N–1 two–stage line evaluations for the throughput and 3N–5 two–stage line evaluations for the acquisition of all average buffer contents. The number of two–stage line evaluations of Glassey and Hong is unpredictable and depends on the initial trial–solution, but the procedure may also fail to converge.

case	v(K)	ϵ	EX_1	ϵ	EX_{N-1}	ϵ	# eval.
10. sim.	0.8366		30.71		22.49		
G&H	0.8356	0.1	32.76	-6.7	22.81	-1.4	44
AP2 (l)	0.8242	1.5			21.16	5.9	2
AP2 (r)	0.8322	0.5	31.06	-1.1			10
11. sim.	0.7314		22.55		35.72		
G&H	0.7280	0.5	22.27	1.2	34.67	2.9	34
AP2 (l)	0.7228	1.2			33.78	5.4	2
AP2 (r)	0.7348	-0.5	21.96	2.6			12
12. sim.	0.7441		28.14		49.64		
G&H	0.7476	-0.5	28.60	-1.6	50.37	-1.5	198
AP2 (l)	0.7362	1.1			40.22	19.0	3
AP2 (r)	0.7294	2.0	29.16	-3.6			3
13. sim.	0.7300		14.68		1.626		
G&H	0.7384	-1.2	15.05	-2.5	1.575	3.1	40
AP2 (l)	0.7284	0.2			1.618	0.5	2
AP2 (r)	0.7298	0.0	14.76	-0.5			2
14. sim.	0.6446		3.649		1.353		
G&H	0.7098	-10.1	3.806	-4.3	1.333	1.5	592
AP2 (l)	0.6516	-1.1			1.313	3.0	4
AP2 (r)	0.6293	2.4	4.071	-11.6			4

Table 4.3.4. Comparison with results of Glassey and Hong

case	N	ν	λ	μ	K	eff
15.	6	(1,1.5,1.2, 0.9,1,1.8)	(.01,.03,.02, .005,.001,.5)	(.09,.05,.08, .07,.009,.5)	(20,20,20, 20,20)	(.9,.94, .96,.84, .9,.9)
16.	7	(1,1,1.2,0.9, 1.5,0.9,1.8)	(.001,.01,.02, (.005,.03,.01, .5)	(.009,.09,.08, .07,.05,.09, .5)	(20,20,20, 20,20,20)	(.9,.9,.96, .84,.94, .81,.9)
17.	8	(1.05,1,1, 1.2,0.9,1.5 0.9,1.8)	(.01,.001,.01, .02,.005,.03, .01,.5)	(.07,.009,.09, .08,.07,.05, .09,.5)	(20,20,20, 20,20,20, 20)	(.92,.9,.9, .96,.84, .94,.81,.9)

Table 4.3.5. Parameters of some long lines

Finally, the approximation method is tested for three longer lines. These lines are given in Table 4.3.5. The results of the approximation can be found in Table 4.3.6. The simulation results were obtained by 10 runs of 20,000 units of time per run. From the results in Table 4.3.6, it is found that method AP2 overestimates the throughput for lines with more than five machines.

case	$v(\mathbf{K})$	ϵ	EX_1	EX_{N-1}	EL	ϵ
15. sim.	0.6302		12.3	5.25	40.2	
N=6 AP2 (l)	0.6469	−2.6		5.98	46.2	−14.9
AP2 (r)	0.6860	−8.9	14.1			
16. sim.	0.6037		17.0	4.15	66.8	
N=7 AP2 (l)	0.6686	−10.8		4.94	70.7	−5.8
AP2 (r)	0.6300	−4.4	17.7			
17. sim.	0.5935		16.1	4.02	81.4	
N=8 AP2 (l)	0.6736	−13.5		5.02	84.9	−4.3
AP2 (r)	0.6394	−7.7	17.3			

Table 4.3.6. Approximation results for the lines in Table 4.3.5

This error can be reduced by setting $K_i'=K_i$ for each fourth aggregation step in such lines. This adaptation of method AP2 is denoted by AP3. The adaptation, which may seem somewhat arbitrary, is only inspired by a number of numeric results and is not supported by analysis. From results in Table 4.3.7 it appears that AP3 performs better for such long lines than AP2. Nevertheless, the errors increase in absolute value if lines become longer.

case	$v(\mathbf{K})$	ϵ	EX_1	EX_{N-1}	EL	ϵ
15. sim.	0.6302		12.3	5.25	40.2	
AP3 (l)	0.6405	−1.6		4.98	46.1	−14.7
AP3 (r)	0.6238	1.0	15.0			
16. sim.	0.6037		17.0	4.15	66.8	
AP3 (l)	0.6230	−3.2		4.22	70.1	−4.9
17. sim.	0.5935		16.1	4.02	81.4	
AP3 (l)	0.6127	−3.2		4.00	84.9	−3.2
AP3 (r)	0.6140	−3.5	17.5			

Table 4.3.7. Approximation results with method AP3 for the lines in Table 4.3.5

It should be remarked that the exact choice of input or output behavior of the two–stage lines in Figs. 4.2.4 and 4.2.5, to base the parameters of the aggregate machines on, is not crucial for the quality of the approximation. From a number of results it appeared that there is generally not much difference in the quality of the approximation when input and output behavior are interchanged.

4.4 OTHER NETWORK LAYOUTS

In this section attention will be paid to two other special network layouts, namely a/d networks (subsection 4.4.1) and buffer–sharing networks (subsection 4.4.2), which were introduced in Chapter 3. It will appear that the approximation method developed in the previous section can be applied straightforwardly to continuous–flow networks with these layouts.

4.4.1 Approximation of assembly–disassembly networks

In this section, an approximation method is developed for continuous–flow a/d networks. Gershwin (1986a, 1986b) also developed an approximation method for a/d networks, with discrete product flow. His method is based on an aggregation algorithm developed for flow lines (Gershwin, 1987). See the description in section 4.3. Liu and Buzacott (1988) give an approximation method for a similar model.

The buffers in the a/d networks are assumed to have finite capacity. The operation of such a/d networks is described in section 3.4. We assume that the networks are tree–shaped. Loops and bypasses are not treated here. It is furthermore assumed that the networks contain a single a/d machine, but it is straightforward to approximate networks with several a/d machines (see De Koster, 1987c). An important tool in the analysis of a/d networks is the reversibility principle, described in section 3.4. For a/d networks this means that reversing the flow direction in one or more buffers of the network (with simultaneous addition of extra input or output streams, if necessary), leads to a sample–path equivalent network. See subsections 3.3.3 and 3.4.1. The approximation method is explained in subsection 4.4.1.1. Numeric results are given in subsection 4.4.1.2. The results are based on De Koster (1987c).

4.4.1.1 networks with a single a/d machine.

In this subsection tree–shaped a/d networks, with a single a/d machine will be considered. In an N–machine a/d system, this single a/d machine is assumed to be machine M_N. In such a network there are N–1 buffers (see Fig. 4.4.1). Each buffer has finite capacity (K_i). The machines have again three parameters (ν,λ,μ). Machine failures are time-dependent.

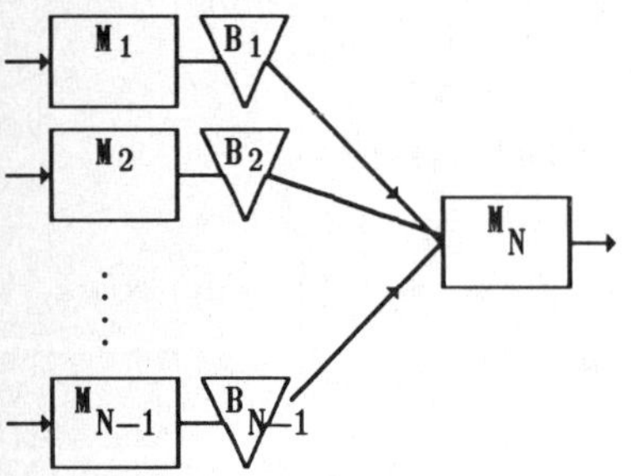

Figure 4.4.1. Assembly network

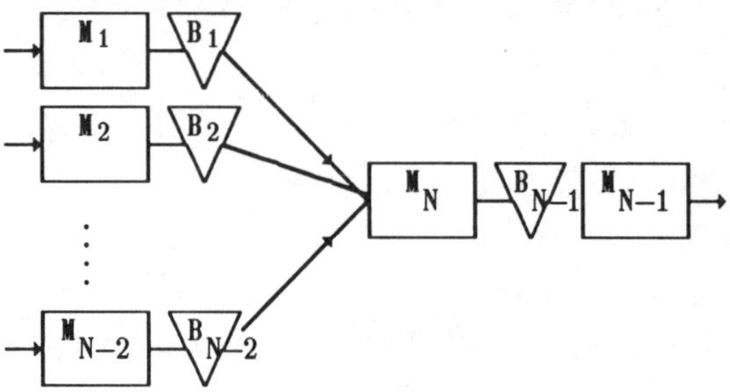

Figure 4.4.2. Equivalent assembly network

The network in Fig. 4.4.2 arises from the network in Fig. 4.4.1 by flow direction reversal in B_{N-1}. It is sample-path-equivalent with the network in Fig. 4.4.1 (see 3.4.1). The throughput of the network in Fig. 4.4.2 equals the throughput of the network in Fig. 4.4.1. Of course it is possible to construct more networks in this way, all equivalent to the network in Fig. 4.4.1.

In approximating a network as in Fig. 4.4.1, first a network is constructed as in Fig. 4.4.2. Then the last two machines are aggregated. Hence a new network is obtained of the type in Fig. 4.4.1, with one machine less. This procedure is repeated until two machines and one buffer are left. The aggregate machine is of the same type as the original machines, that is, with an exponentially distributed life and repair time. In each aggregation step, the input or the output behavior of the two-stage line to be approximated, may be used to determine the three parameters (ν,λ,μ) of the aggregate machine. The determination of the parameters of the aggregate machine is as described in section 4.1. The approximation method is similar to method AP2 for flow lines.

Assume that the behavior of the contents of B_1 and the throughput of the system sketched in Fig. 4.4.1, is to be approximated. A possible sequence for carrying out the aggregation is depicted in Fig. 4.4.3. The approximation consists of N–1 steps. In aggregation step i ($1 \leq i \leq N-2$), the flow in buffer B_{N-i} is reversed. Its capacity is adapted, in a similar way as in the scheme in Fig. 4.2.5. That is,

$$K_i' = K_i + E\tilde{X}_{i+1} \quad , 2\leq i\leq N-2$$
$$K_1' = K_1$$
$$K_{N-1}' = K_{N-1}. \tag{4.1}$$

The quantity $\tilde{X}_j$ is defined as the buffer content of the corresponding two–stage line in the (N–j)th step. Then the two–stage line agg_{i-1}–B_{N-i}–M_{N-i} is aggregated. The aggregate agg_{i-1} is obtained in the (i–1)st step ($agg_0:=M_N$). The aggregation is carried out with respect to output behavior. The average amount of products in agg_{i-1} is $E\tilde{X}_{N-i+1}$, which supports the definition of K_{N-i} in (4.1).
In the (N–2)nd aggregation step the input behavior of the two–stage line is taken, since eventually the behavior of the contents of B_1 is desired. In step N–1 the behavior of the content of B_1 and the throughput estimate of the original network are determined. Note that agg_{N-2} actually represents the whole network "downstream" of B_1.

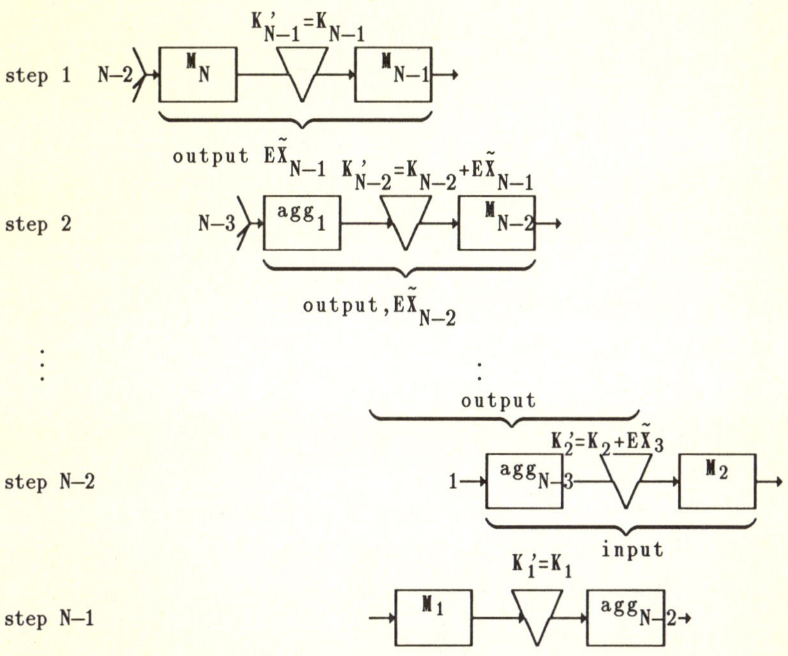

Figure 4.4.3. Approximation of the network in Fig. 4.4.1 (throughput and contents of B_1).

If the behavior of the contents of another buffer has to be estimated, then the aggregation sequence can be adjusted so that this buffer remains in the last step. If only the throughput of the system is desired, then a single aggregation sequence suffices. Note that, similarly as for flow lines, all aggregation sequences yield different throughput estimates. The average value may serve as ultimate estimate.

The approximation of a/d networks with a single a/d machine and other layouts than sketched in Fig. 4.4.1 is similar to the scheme in Fig 4.4.3. The approximation of the behavior of the content of B_4 in the network in Fig 4.4.4 is treated as an example.

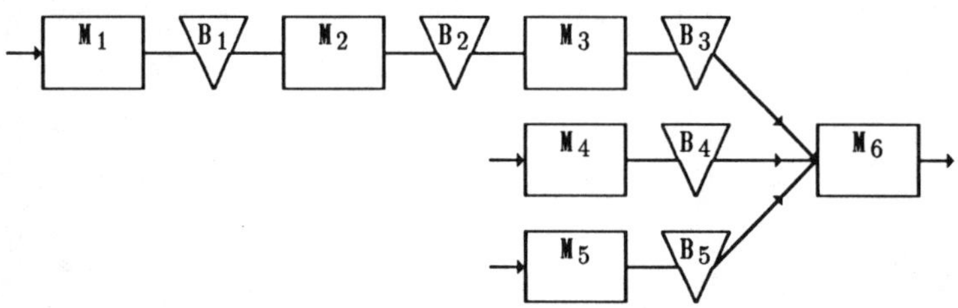

Figure 4.4.4. 6–machine a/d network

To approximate the behavior of the mean contents of B_4 and the throughput for the network in Fig. 4.4.4 at the same time, we can aggregate as follows.

Step 1. Reverse the flow in B_5. Aggregate over M_6–B_5–M_5 with respect to the output behavior. Let agg_1 be the aggregate machine. By reversing the flow direction in B_4 we obtain the flowline M_1–B_1–M_2–B_2–M_3–B_3–agg_1–B_4–M_4. It should be noted that the mean contents of B_4 in this resulting line approximates K_4 minus the mean contents of B_4 in the original system, since the flow in B_4 is reversed.

Step 2. The flow line obtained now can be approximated from left to right by method AP2, indicated in Fig. 4.2.4, in order to obtain an approximation of the behavior of the content of B_4.

From the examples given above, it will be clear that the approximation method works for arbitrary tree–shaped a/d networks. In De Koster (1987c) a number of examples are given for the following cases:

1. a/d networks with more than one a/d machine.

2. a/d networks with different assembly/disassembly ratios for the different branches.
3. Some special a/d networks with loops and bypasses.

4.4.1.2 Numeric results

In this subsection, some numerical results of the approximation method will be presented. The approximated throughput and mean buffer contents are compared with simulation results. We consider a four–, a five– and a six–machine assembly network with layouts as in Fig. 4.4.1. Furthermore, a network with a layout as in Fig 4.4.4 is considered. The reason that no three–machine assembly (or dis–assembly) networks are discussed, is that each three–machine assembly or dis–assembly network is equivalent to an ordinary flow line (see subsection 3.3.3). The first network to consider is a four–machine network, of which the parameters and the approximation results are given in Table 4.4.1. The approximation results were derived for varying buffer capacities, by the scheme indicated in Fig. 4.4.3. The simulated mean buffer content of buffer B_i is denoted by EX_i and the throughput by $v(\mathbf{K})$. The corresponding results of the approximation method are denoted by the index ap. Relative errors (in %) in both throughput and mean buffer contents compared with the simulation results are denoted by ϵ. In Table 4.4.2, parameters and results for a five–machine network are listed.

K_1	K_2	K_3	$v(\mathbf{K})$	$v_{ap}(K_1)$	ϵ	EX_1	$EX_{1,ap}$	ϵ
2	2	2	.6169	.6104	1.1	1.66	1.72	–3.6
5	5	5	.6593	.6533	0.9	3.95	4.17	–5.6
10	10	10	.7030	.6986	0.6	7.64	8.09	–5.9
10	12	15	.7178	.7124	0.8	7.49	8.02	–7.1
15	20	25	.7502	.7437	0.9	10.96	11.79	–7.6
30	30	30	.7737	.7712	0.3	22.54	23.65	–4.9

Table 4.4.1. Approximation of a four–stage a/d network. $\boldsymbol{\nu}$=(1.05,0.95,0.9,1) $\boldsymbol{\lambda}$=(0.01,0.01,0.01,0.01), $\boldsymbol{\mu}$=(0.07,0.08,0.09,0.09), **eff**=(0.919,0.844,0.81,0.9).

Recall that in the case of flow lines, the performance of the approximation method deteriorates with the line length. For a/d networks, there are similar difficulties if the networks become larger. For networks consisting of five or more buffers, we use an analogue of method AP3 (see Table 4.3.7). This adapted method is tested on a six-stage a/d network. The parameters and approximation results are given in Table 4.4.3. The buffer capacity in the fourth approximation step is unadapted, or $K_2' = K_2$.

K_1	K_2	K_3	K_4	$v(\mathbf{K})$	$v_{ap}(K_1)$	ϵ	EX_1	$EX_{1,ap}$	ϵ
3	3	3	3	.5705	.5634	1.2	2.60	2.69	−3.5
3	4	5	6	.5979	.5904	1.3	2.46	2.58	−4.9
10	10	10	10	.6587	.6547	0.6	7.97	8.32	−4.4
15	12	12	15	.6841	.6834	0.1	11.93	12.36	−3.6
20	20	20	20	.7144	.7115	0.4	15.72	16.34	−3.9

Table 4.4.2. Approximation of a five-stage a/d network. ν=(1.05,0.95,0.9,0.85,1) λ=0.01,0.01,0.01,0.01,0.01), μ=(0.07,0.08,0.09,0.1,0.09), eff=(0.919,0.844,0.81,0.765,0.9).

K_1	K_2	K_3	K_4	K_5	$v(\mathbf{K})$	$v_{ap}(K_1)$	ϵ	EX_1	$EX_{1,ap}$	ϵ
5	5	5	5	5	.5798	.5691	1.8	4.13	4.24	−2.7
10	10	10	10	10	.6541	.6422	1.8	7.96	8.17	−2.6
20	25	25	25	25	.7533	.7384	2.0	19.00	19.50	−2.6

Table 4.4.3. Approximation of six-stage a/d network. ν=(1.05,1.1,1.2,0.9,1,1), λ=(0.01,0.02, 0.01,0.01,0.01,0.01), μ=(0.07,0.09,0.08,0.1,0.09,0.09), eff=(0.919,0.9,1.067,0.818,0.9,0.9).

In all three networks the relative error in throughput is below 2%, the relative error in mean content of B_1 is below 8%.

Results of the approximation of the throughput and the content of B_4 for a network with layout as in Fig. 4.4.4 are given in Table 4.4.4.

					simulation		approximation	
K_1	K_2	K_3	K_4	K_5	$v(\mathbf{K})$	EX_4	$v_{ap}(K_4)$	$EX_{4,ap}$
5	5	10	10	10	.5879	8.06	.6011	8.64
15	15	15	10	10	.6548	7.58	.6777	8.29

Table 4.4.4. Approximation results for a network with layout as in Fig. 4.4.4. ν=(1,1.1,1.2, 1,0.9,1.05), λ=(0.01,0.02,0.03,0.01,0.01,0.01), μ=(0.09,0.09,0.09,0.09,0.1,0.07), eff=(0.9,0.9,0.9,0.9, 0.82,0.919).

4.4.2 Buffer–sharing networks

In this subsection, the approximation method developed in sections 4.1 and 4.2 will be applied to buffer–sharing networks, as introduced in Chapter 3. See Fig. 3.4.4. A speed allocation rule has to be specified for buffer–sharing networks, that allocates the speed reduction to the machines in case a common downstream buffer is full, or a common upstream buffer is empty. Different allocation rules may lead to a different system performance. The approximation of a buffer–sharing network will be explained by means of an example network, sketched in Fig. 4.4.5. All machines have exponential life and repair times and a constant machine speed. The machine parameters (ν,λ,μ) and the machine efficiencies are indicated in Fig. 4.4.5.

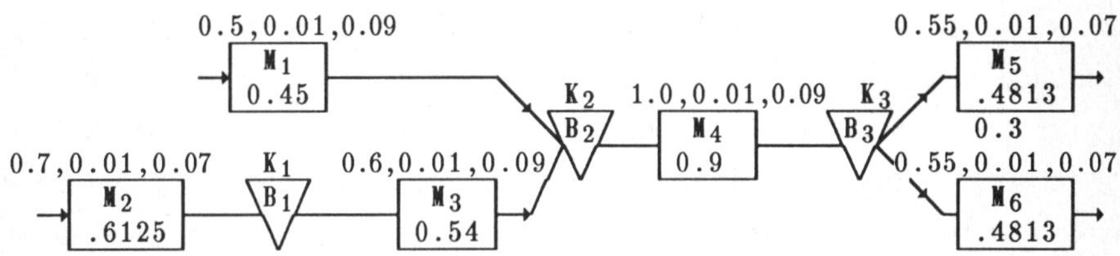

Figure 4.4.5. Buffer–sharing production system

For the system in Fig. 4.4.5 it is assumed that the allocation rule is rule R_1, introduced in subsection 3.4.2. Note that allocation rule R_2, also defined in sub–section 3.4.2, would yield a different throughput, because the behavior of buffer B_1 is influenced. The approximation method cannot discriminate between different allocation rules. A possible approximation sequence for the system is the following.

step 1. The two–stage system M_2–B_1–M_3 is aggregated. Call the aggregate agg_1.

step 2. The parallel machines M_1 and ag_1 are combined to form a four–state machine M_7, as indicated in subsection 2.4.2. In the same way the parallel machines M_5 and M_6 are combined to a three–state (namely the number of "up" machines, see subsection 2.4.2) machine M_8. The resulting system is a three–stage line M_7–B_2–M_4–B_3–M_8.

step 3. This flow line can be approximated using method AP2 (see section 4.2).

Note that the flow line in step 3 consists of multistate machines. Therefore, in approximating two machines and an intermediate buffer, the aggregation method for multistate machines described in section 4.1b has to be used. In general, this aggregation method for multistate machines has the drawback that it is only applicable if the number of machine states is not too large. To be able to tackle arbitrary buffer–sharing networks, we propose to use first the approximation method for complex two–stage lines of section 2.7. This intermediate approximation step reduces the number of machine states of a two–stage line, modeled as in section 2.4, without affecting the accuracy much. The approximating two–stage line has only four states (see Fig. 2.7.2). Its parameters are determined by maintaining equal the first three moments of stock increase and stock decrease in both lines. Furthermore, the parameters are so chosen that the expected lengths of stock increase and stock decrease periods are equal in the two lines, and the throughputs are equal for zero and infinite buffer capacities (relations (2.60), (2.64) and (2.65) hold). After this "state–reduction step", as we will call it from now on, the aggregation step for multistate machines of section 4.1b can be applied without computational difficulties.

Another problem in step 3 is, whether or not the buffer capacities have to be adapted in an approximation from left to right (since we deal with a three–stage line) in order to obtain a throughput estimate. This depends on whether or not machine M_4 is faster on average than machine M_7, in conformity with method AP2. However, for all investigated values of K_1, $eff_7 > eff_4$, so that the buffer capacities need not be adapted. In step 1 we therefore use the output behavior to obtain agg_1. Note that $eff_4 < eff_8$, so that the buffer capacities also need not be adapted in approximation from right to left. The results of the approximation of the system are given in Table 4.4.5.

				left		right	
K_1	K_2	K_3	$v(K)$	$v_{ap}(K_3)$	rel.err.(%)	$v_{ap}(K_2)$	rel.err.
4	6	6	0.7983	0.8006	–0.29	0.8028	–0.56
5	10	10	0.8246	0.8262	–0.19	0.8278	–0.39
10	10	10	0.8311	0.8331	–0.24	0.8330	–0.23
10	15	20	0.8558	0.8575	–0.20	0.8566	–0.09
10	20	15	0.8551	0.8565	–0.16	0.8568	–0.20

Table 4.4.5. Approximation of the throughput of the system in Fig.4.4.5, from left to right and from right to left

From results in Table 4.4.5 we see that the approximation performs well both from left to right and from right to left. Relative errors are in all cases below 1%.

From this example, it will be clear that approximation of buffer–sharing networks is in principle as simple as the approximation of flow lines. In De Koster (1988a) several other examples are given, for networks up to nine machines and four buffers. One of the examples contains a bypass. The results are found to be best when the networks show some resemblance to a flow line. This is due to the fact that all networks are simulated with allocation rule R_1. This allocation rule discriminates between different branches that supply products to a common buffer, or different branches that use products from a common buffer. The approximation method cannot do that. However, if the layout is that of a flow line, with possible parallel machines, then allocation rules R_1 and R_2 will yield the same throughput (see also subsection 3.4.2).

It will also be clear that the assumption of two–state machines in the system in Fig. 4.4.5 is not important. For the case of multistate machines, modeled as in section 2.4, it is possible to apply the approximation method of section 2.7 in all kinds of layout such as flow lines, a/d networks or buffer–sharing networks, before applying the aggregation method of section 4.1. This state reduction step will hardly affect the quality of the approximation.

4.5 OTHER TYPES OF LOCAL CONTROL

This section deals with lines controlled with $<m,M>$ *switching–point policies* as described in subsection 2.5.2. As soon as the inventory drops below level m, the preceding stage tries to replenish it, until level M is reached, after which it stops producing.

Subsection 2.5.2 described how two–stage lines controlled by such an $<m,M>$ policy can be analyzed. The results of that section can be combined with the aggregation method of section 4.1, to approximate the behavior of larger continuous–product flow lines controlled by such $<m,M>$ policies. Some flow lines with a layout as sketched in Fig. 4.5.1 will be approximated as an example.

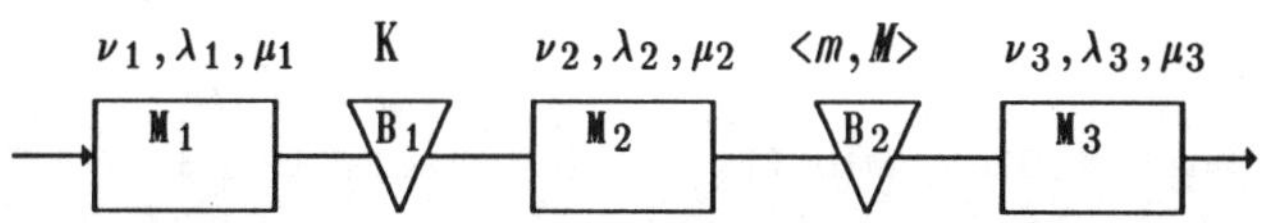

Figure 4.5.1. Locally controlled flow line, with $<m,M>$ control for B_2

The machines are unreliable at speed ν, failure rate λ and repair rate μ. The second buffer in Fig. 4.5.1 has $<m,M>$ control, the first buffer is finite with capacity K.

To approximate the line in Fig. 4.5.1 for the throughput, from left to right, two cases have to be distinguished.

1. $eff_1 < eff_2$ or $eff_1 = eff_2$ and M_2 is more stable than M_1.

In this case M_1–B_1–M_2 has to be aggregated with respect to input behavior, in the first step. This results in an aggregate machine agg. In the second step, the line agg–B–M_3, with B a buffer controlled by a $<m,M+K_1-E\tilde{X}_1>$ switching–point policy, is evaluated for the throughput. $E\tilde{X}_1$ is the expected buffer content of the line aggregated in the first step.

2. $eff_1 > eff_2$.

In this case M_1–B_1–M_2 has to be aggregated with respect to output behavior, which results in an aggregate machine, denoted by agg. Then throughput of the line agg–B_2–M_3, with B_2 controlled by an $<m,M>$ switching policy.

If the line consisted of more than three stages, then the throughput could be estimated by adapting the upper switching points of all but the first and the last buffers, in accordance with method AP2. The procedure conforms to the approximation method AP2 described in section 4.2.

The parameters of the lines to be approximated are given in Table 4.5.1.

case	ν	λ	μ	eff	K	m	M
1.	(1.05,1,1.6)	(.01,.01,.5)	(.07,.09,.5)	(,92,.9,.8)	10	10	20
2.	(0.95,1,1.6)	(.01,.01,.5)	(.09,.08,.5)	(.86,.89,.8)	10	10	20
3.	1,1.5,1.8)	(.01,.01,.005)	(.1,.09,.005)	(.91,1.35,.9)	10	10	20

Table 4.5.1. Parameters of lines with layouts as in Fig. 4.5.1

Approximation results are listed in Table 4.5.2. The simulated throughput was obtained by 10 runs of 20,000 units of time. It will be clear that this type of control can be extended to all network layouts studied in the previous sections. The approximation method is still applicable and the results in Table 4.5.2 suggest that they will be reasonable.

case	simulation	approximation	error (%)
1.	0.7308	0.7244	0.9
2.	0.7289	0.7316	−0.4
3.	0.4851	0.5036	−3.8

Table 4.5.2. Approximations of lines given in Table 4.5.1

4.6 DISCUSSION

In this chapter a method was described for the evaluation of system quantities, such as throughput and mean buffer content.

1. The method is fast: an aggregation step as described in section 4.1 takes about one second of CPU time on a Burroughs B7900 computer. The exact time depends on the method used, on the required accuracy and on the buffer capacity (see section 2.3.5).
2. The approximation method described is reasonably accurate for systems up to four buffers. For larger systems the errors increase, but as a fast tool the method is still satisfactory.
3. The method is reasonably generally applicable and different layouts can be approximated by means of it. In the previous sections some examples of applications are given, but it will be clear that other layouts can also be dealt with: combined buffer–sharing–a/d networks, networks with loops, buffer–sharing. Also buffer–sharing networks with other priority rules can be dealt with. It may be, however, that the approximate performance becomes less accurate for more complicated systems.

The approximation method can be used to establish norms for the capacities. By evaluating system performance (for instance system throughput) for varying buffer capacities, the appropriate buffer capacities can be chosen.

CHAPTER 5

ANALYSIS OF INTEGRALLY CONTROLLED PRODUCTION SYSTEMS

The production systems dealt with in the previous chapters were locally controlled, either by finite buffers (Chapters 2, 3 and 4) or by $<m,M>$ switching policies (sections 2.5.2 and 4.5). In locally controlled systems, the speed of a machine is directly influenced by those of neighboring machines. In real production systems, besides local dependencies between machines, integral dependencies can also be observed. For example, in Flexible Manufacturing Systems (FMS), the machines may have a common storage (Buzacott and Shantikumar, 1980). One way of realizing such a common storage space is sketched in Fig. 5.1.

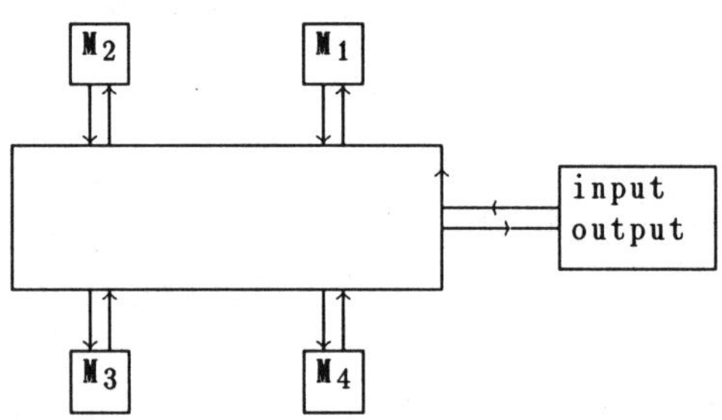

Figure 5.1. FMS with loop conveyor

The loop conveyor in Fig. 5.1 acts as both a common storage and a transport device for machines 1–4. The different machines may also have additional local storage. Here the finite buffer space is caused by a physical restriction on the size of the conveyor. In such an FMS, the routing of the products over the machines may be variable. However, even in case of a fixed routing of the products, a loop conveyor may have an advantage over a flow line, since the input is not blocked before the conveyor is filled completely. In this example, the integral buffer restriction has a physical origin. The number of products in the loop is limited. Buffer restrictions may also have a nonphysical nature. They may also be caused by the way the system is controlled. For example, they may be software implemented in highly automated FMS systems. Other examples of integral restrictions caused by the control of the system are given by Buzacott (1988). He demonstrates that certain Kanban and MRP controlled–production systems can be modeled as having restrictions on the number of products in (groups of) adjacent stages.

In this chapter, integral control structures for single product flow lines are studied. Two specific control structures are considered for flow lines, namely the Base–Stock System (BSS) and the Workload Control system (WC), introduced in Chapter 1. These control structures cover loop conveyors with fixed routings for the products. It turns out that the performance of an integrally controlled (WC or BSS) single–product flow line, with continuous or discrete product flow, can also be (approximately) obtained by applying local control by finite buffers, provided the buffer capacities are chosen properly. The performance of a line is measured in the sequel by the combination of throughput v and average total system content EL.

The possibility to transfer these results to multiproduct lines and to more general production stages will be discussed in Chapter 6.

In subsection 5.1.1 the working of the Base–Stock System is described for a single–product continuous flow line with unreliable machines. For such lines controlled with BSS, a method is given in subsection 5.1.2, to obtain the buffer capacities of a locally controlled line with approximately the same performance as the original line. This method is based on the approximation method of section 4.1. In that section numerical results are also given. Results in this section are based on De Koster (1986a).

It is demonstrated in section 5.2 that workload–controlled (continuous product) flow lines can also be approximated by locally controlled ones. Simulations are used to demonstrate the qualities of the approximations. The equivalence of local and integral control is not due to the fact that the product flow is continuous. This is argued in section 5.3, where examples are given that show that similar results hold for a specific discrete product flow–line model. The results of sections 5.2 and 5.3 are based on De Koster and Wijngaard (1987a). Finally, the results of this chapter are discussed in section 5.4.

5.1 CONTROL WITH THE BASE–STOCK SYSTEM

5.1.1 Model description

The Base–Stock System (BSS) was introduced in Chapter 1. Recall that BSS relies on the notion of echelon stocks, i.e. the stock in the system downstream of the stages. Here it will be applied to continuous–product flow lines as sketched in Fig. 5.1.1.

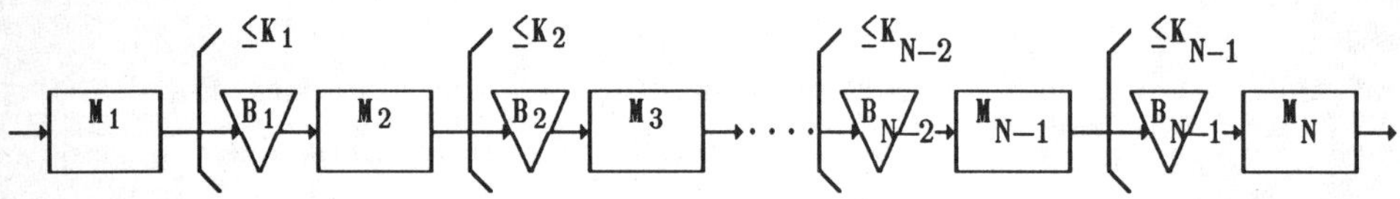

Figure 5.1.1. Flow line controlled by BSS

Machine M_i in Fig. 5.1.1 produces at its maximum rate if the inventory in (right) echelon i drops below the level K_i. Right echelon i is the part of the line downstream of machine M_i. Hence K_i is a limit on the total inventory and work–in–process in the part downstream of M_i. Machine M_i is blocked if the right echelon i is full and machine M_N is down. It may also be starved or slowed down. See subsection 1.3.2. The reason for controlling a line by BSS, is to load machine M_N as efficiently as possible: all other machines are allowed to produce as soon as an item leaves the system. In this way, a high throughput can be realized with little work–in–process.

The Base–Stock System is usually not applied to machine lines like the one sketched above. The BSS is intended to coordinate departments. Machine M_N should then be interpreted as the demand by customers as experienced at stockpoint B_{N-1} and starvation of M_N should be interpreted as inability to satisfy demand. Here it is assumed that control by BSS is also meaningful for simple lines like the one sketched in Fig. 5.1.1. The system is sensible anyway if M_N is the only bottleneck. Since all machines are tuned on M_N, which eventually determines the throughput of the line, one may conjecture control by BSS to lead to better system performance than control by finite buffers.

In subsection 5.1.2 it will be shown that this conjecture is false. An approximation technique is presented to determine the buffer capacities in locally controlled lines that have the same performance. Such a locally controlled line consists of the same

machines as the BSS–controlled line. This technique used to determine local buffers is similar to the approximation method of Chapter 4. It can also be used for approximating the performance of BSS–controlled lines.
To avoid trivialities, it is assumed that the echelon capacities are such that $K_1 \geq K_2 \geq \ldots \geq K_{N-1}$. It is assumed that the physical capacity of B_i is so large that no blocking occurs before echelon i is full. Hence $cap(B_i) > K_i$, for all i.

5.1.2 Approximation by locally controlled lines.

It is assumed that the machine model is the same as in the previous chapter: each machine has three parameters, a speed, failure and repair rate (ν,λ,μ), with exponentially distributed life and repair times. Machine failures are time–dependent.

The line sketched in Fig. 5.1.1 is approximated by a locally controlled line (as sketched in Fig. 4.2.1) consisting of the same machines, but with local buffer capacities K_i'. The buffer capacity K_i' equals K_i minus the approximate average inventory downstream of M_{i+1}. The average inventory downstream of M_{i+1} is determined by an approximation scheme similar to the scheme in Fig. 4.2.5. The scheme consists of N–2 steps and is clarified in Fig. 5.1.2.

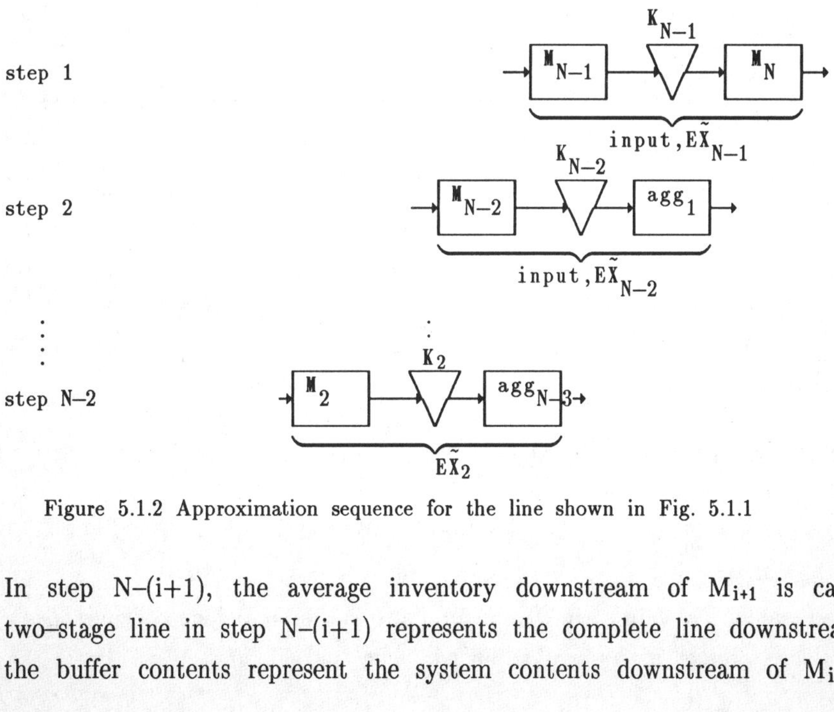

Figure 5.1.2 Approximation sequence for the line shown in Fig. 5.1.1

In step N–(i+1), the average inventory downstream of M_{i+1} is calculated. The two–stage line in step N–(i+1) represents the complete line downstream of B_i and the buffer contents represent the system contents downstream of M_{i+1}. The mean

buffer content is denoted by $E\tilde{X}_{i+1}$. In step N–(i+1) the two–stage line is furthermore approximated by an aggregate machine, denoted by agg_i, using the aggregation method described in section 4.1.
Hence, the local buffer capacities K_i' are determined as

$$K_i' = K_i - E\tilde{X}_{i+1} \quad , i = 1,\dots,N-2$$
$$K_{N-1}' = K_{N-1} \tag{5.1}$$

Note the possibility of $K_i' \simeq 0$, namely in the case of $K_i \simeq K_{i+1}$ and $E\tilde{X}_{i+1} \simeq K_{i+1}$. This, however, does not cause problems for the approximation procedure.
The scheme in Fig. 5.1.2 differs from the scheme in Fig. 4.2.5. In all steps of the scheme in Fig 5.1.2 the input behavior of the two–stage lines is used in the aggregation, while in the scheme in Fig 4.2.5 output behavior is used to include inventory in aggregates. As remarked in section 4.3, the use of input and output behavior is not really important, since it does not lead to great differences in performance.

Once the K_i' have been determined, the throughput of the local buffer line can be approximated by the approximation technique AP2 described in section 4.2. In Table 5.1.1 four lines are given for which control by BSS will be compared with local control. More examples can be found in De Koster (1986a).

case	N	ν	λ	μ	eff
1.	3	(1,0.9,1)	(0.01,0.01,0.01)	(0.09,0.07,0.09)	(0.9,0.7875,0.9)
2.	3	(0.9,10,1)	(0,01,0.01,0.01)	(0.09,0.09,0.09)	(0.81,9,0.9)
3.	4	(10,10,10, 1)	(0.01,0.01,0.01, 0.01)	(0.09,0.09,0.09, 0.09)	(9,9,9,0.9)
4.	5	(0.9,1.05,1, 0.9,1)	(0.01,0.01,0.01, 0.01,0.01)	(0.09,0.07,0.09, 0.07,0.09)	(0.81,0.919,0.9, 0.7875,0.9)

Table 5.1.1. Parameters of four different lines

The parameter choices are motivated as follows.
1. The number of machines varies from 3 to 5.
2. The repair rates are greater than the failure rates. Failure rates are all chosen as 0.01.
3. Balance vs unbalance. All lines in Table 5.1.1 are rather unbalanced, apart from the fourth one, which is almost balanced. The second and third lines are

extremely unbalanced. The third line is a situation typical for control by BSS. M_4 can be interpreted as demand, which has to be satisfied by machines M_1, M_2 and M_3 that have overcapacity compared with M_4.

4. Location of the bottleneck and fastest machine. In cases 1 and 4, the bottleneck is in the middle of the line. In case 2 the fastest machine is in the middle, the bottleneck is at the front of the line. In case 3 the bottleneck is at the end of the line.

The base–stock–controlled line and the corresponding locally–controlled line are compared in performance by simulation, for different values of the echelon capacities. All simulation results in the rest of this section and in section 5.2 are obtained by 10 runs of 500,000 units of time per run on a Burroughs B7900 computer. By applying the aggregation technique all K_i' (i=1,..,N–2) are determined in N–2 steps (note that $K_{N-1}' = K_{N-1}$). The approximating local buffer line was simulated using the same random numbers as the BSS line. Results for the first line are given in Table 5.1.2.

K_1	K_2	K_1'	v(K)	v_{lc}(K')	ε(%)	EL	EL_{lc}	ε(%)	$\hat{v}(K_1')$
5	2	4.71	.6771(7)	.6770(8)	0.0	4.16(0)	4.19(0)	–0.7	.6768
10	4	9.28	.7008(7)	.7010(8)	–0.0	8.11(1)	8.19(1)	–1.0	.7008
20	8	18.29	.7295(7)	,7300(7)	–0.1	16.13(2)	16.30(2)	–1.1	.7298
30	12	27.29	.7463(7)	.7469(7)	–0.1	24.46(4)	24.75(4)	–1.2	.7468
50	20	45.49	.7646(7)	.7650(7)	–0.1	41.99(4)	42.49(8)	–1.2	.7652
20	20	15.49	.7442(7)	.7466(7)	–0.3	15.91(2)	16.18(3)	–1.7	.7460

Table 5.1.2. Approximations of the first line in Table 5.1.1

EL denotes the simulated average system contents (BSS) and v(K) the throughput of the BSS line. The corresponding quantities for the locally–controlled line are indicated by subscript lc. The notation .6771(7) for v(K), $v_{lc}(K')$, EL and EL_{lc} means that the 95% confidence interval of the simulated value is given by (0.6764, 0.6778). The relative difference (ε) of v(K) and $v_{lc}(K')$ is less than 0.1%, except for the case of K_1=K_2=20. The line with K_1=K_2=20 is special. It can be interpreted as a line with a single, common finite capacity buffer for all machines. It is equivalent to a closed–loop production line as sketched in Fig. 5.1.3. This can be seen easily by comparing blocked, starved and slowed–down states of the machines in the systems.

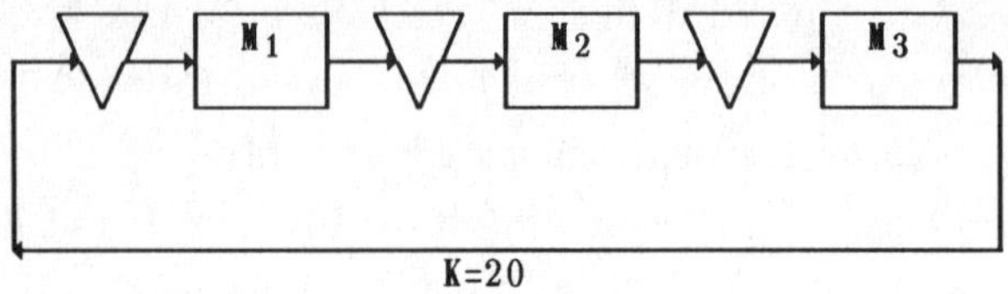

Figure 5.1.3. Closed–loop system, containing 20 products and without buffer restrictions

There is another method for obtaining a throughput estimate for the BSS line, besides the approximation of a BSS line by a finite buffer line, and then subsequently applying method AP2. That is by applying only one additional aggregation step in the scheme in Fig. 5.1.2 and calculating the throughput of the line M_1–B–agg_{N-2}, where B is a buffer of capacity $K_1' = K_1 - E\tilde{X}_2$. This method for obtaining a throughput estimate for the BSS line resembles the method AP2 of Fig 4.2.5 for locally controlled lines (buffers K_i'), except for the fact that input instead of output behavior of two–stage machine lines is used. The throughput estimate obtained by applying this extra step is denoted by $\hat{v}(K_1')$. Results for this approximation method are also indicated in Table 5.1.2. Note that this approximation method performs good (relative errors are smaller than 1%).

Approximation results for the second line are given in Table 5.1.3.

K_1	K_2	K_1'	$v(K)$	$v_{lc}(K')$	$\epsilon(\%)$	EL	EL_{lc}	$\epsilon(\%)$
5	2	3.17	.7068	.7007	0.9	1.74	1.40	19.5
10	4	6.32	.7352	.7274	1.1	3.93	3.20	18.6
20	8	12.55	.7681	.7602	1.0	7.96	6.67	16.2
30	12	18.72	.7857	.7792	0.8	11.21	9.69	13.6
50	20	30.91	.8017	.7982	0.4	15.55	14.20	8.7
20	20	0.91	.7735	.7222	6.6	7.70	3.80	50.6

Table 5.1.3. Approximations of the second line in Table 5.1.1

The results of the approximation $v_{lc}(K')$ in this table are not of the same quality as those in Table 5.1.2. This is due to the high unbalance of the line. M_2 is, on average, 10 times faster than the other machines. Considerable errors arise, especially in the case of $K_1 \simeq K_2$. Here $E\tilde{X}_2 \simeq K_2$ and hence $K_1' = K_1 - E\tilde{X}_2 \simeq 0$. Moreover, also in the case $K_1 \neq K_2$, K_1' is too small and hence $v_{lc}(K')$ is smaller

than v(**K**) and EL_{lc} is smaller than EL. However, except for the case (K_1,K_2) = (20,20), the maximum relative error in $v_{lc}(\mathbf{K}')$ is still about 1%.
Although situations exist in which the method for obtaining the K_i' does not perform well, it is still possible to obtain a good approximating locally controlled line. This can be done by increasing the K_i' obtained. This leads to a system with higher system contents and higher throughput. As an example this method is applied to the case $K_1=K_2=20$ which has the greatest relative differences. Results are given in Table 5.1.4.

K_1	K_2	K_1'	v(**K**)	$v_{lc}(\mathbf{K}')$	ε(%)	EL	EL_{lc}	ε(%)
20	20	10.45	.7735	.7711	0.3	7.70	7.87	−2.2

Table 5.1.4. Approximation of the second line in Table 5.1.1 for $K_1=K_2=20$

It turns out from Table 5.1.4, that increasing K_1' from 0.91 to 10.45 decreases the relative differences in both throughput and average system contents substantially. The same can be done for the other cases listed in Table 5.1.3.

The results for the third and fourth lines are listed in Tables 5.1.5 and 5.1.6, respectively.

K_1	K_2	K_3	K_1'	K_2'	v(**K**)	$v_{lc}(\mathbf{K}')$	ε(%)	EL	EL_{lc}	ε(%)
5	3	2	2.24	1.17	.7165	.7175	−0.1	4.64	4.63	0.2
10	6	4	4.43	2.32	.7606	.7616	−0.1	9.41	9.39	0.2
20	12	8	8.69	4.55	.8167	.8169	−0.0	19.13	19.10	0.2
30	18	12	12.85	6.72	.8486	.8483	0.0	29.00	28.96	0.1
50	30	20	21.01	10.91	.8792	.8788	0.0	48.90	48.85	0.1
20	20	20	0.89	0.91	.8551	.8537	0.2	19.10	19.01	0.5

Table 5.1.5. Approximations of the third line in Table 5.1.1

From Tables 5.1.5 and 5.1.6 we see that the relative differences increase in longer lines when more aggregation steps are involved. However, all relative errors, except for the case of $K_1=K_2=K_3=K_4=40$ in $v_{lc}(K')$ are less than 1%.
Note that the relative difference in average system contents in the two lines is quite large in Table 5.1.6, for the case $K_1=K_2=K_3=K_4$. This relative difference can be reduced by increasing K_1', K_2' and K_3'. Results are given in Table 5.1.7.

K_1	K_2	K_3	K_4	K_1'	K_2'	K_3'	$v(\mathbf{K})$	$v_{lc}(\mathbf{K}')$	$\epsilon(\%)$	EL	EL_{lc}	$\epsilon(\%)$
5	4	3	2	1.62	1.45	2.71	.5599	.5562	0.7	3.85	3.41	11.4
10	8	6	4	3.41	3.07	5.28	.5986	.5944	0.7	7.72	6.99	9.5
20	16	12	8	7.09	6.38	10.29	.6496	.6449	0.7	15.42	14.20	7.9
30	24	18	12	10.71	9.62	15.29	.6824	.6773	0.7	23.03	21.30	7.5
50	40	30	20	17.54	15.71	25.49	.7219	.7167	0.7	37.97	35.04	7.7
40	40	40	40	8.15	7.35	32.37	.7149	.6988	2.3	29.88	24.37	18.4

Table 5.1.6. Approximations of the fourth line in Table 5.1.1

K_1	K_2	K_3	K_4	K_1'	K_2'	K_3'	$v(\mathbf{K})$	$v_{lc}(\mathbf{K}')$	$\epsilon(\%)$	EL	EL_{lc}	$\epsilon(\%)$
40	40	40	40	11	10	34	.7149	.7129	0.3	29.88	29.41	1.6

Table 5.1.7. Approximations of the fourth line in Table 5.1.1 for the case K_i=40

Even when the approximation method for obtaining the K_i' does not lead to good results, it is still possible to obtain an equivalent locally controlled line, by adapting the K_i'. This is, because the relative differences in throughput and average system contents have the same sign. If the differences are positive, then the K_i' have to be increased, if they are negative, the K_i' have to be decreased. See Tables 5.1.4 and 5.1.7..

5.1.3 Additional local restrictions

The same approximation method can be applied to flow lines controlled by BSS, combined with local finite–capacity buffers. Assume that the physical capacity of buffer B_i is C_i, with $C_i \leq K_i$ and $\sum_{j=i}^{N-1} C_j \geq K_i$ (i=1,..,N–1). Note that the restriction set by K_i becomes ineffective for M_i if $\sum_{j=i}^{N-1} C_j < K_i$. In the case of $\sum_{j=i}^{N-1} C_j < K_i$, for all i, the line reduces to an ordinary flow line where all machines are controlled by finite local buffers. If both local and global restrictions are active, machines may be blocked because the next echelon is full, but also because the next buffer is full.

In the approximation of such a line we proceed as in subsection 5.1.2. However, the K_i in Fig. 5.1.2 have to be replaced by C_i and the K_i' are defined as

$$K_i' = \min\{C_i, K_i - E\tilde{X}_{i+1}\}, \; i=1,2,..,N-2$$
$$K_{N-1}' = C_{N-1} \tag{5.2}$$

Results are given in Table 5.1.8 for a four–machine line with physical buffer capacities $C_i \leq K_i$. C_{N-1} equals K_{N-1}. Although relative differences in the average line contents increase considerably with additional local buffers, the relative errors in throughput are still slight. Furthermore, the relative errors in both throughput and average system contents can be simultaneously reduced by adapting (i.e. decreasing) K_1' and K_2', similarly as before.

K_1	K_2	K_3	C_1	C_2	K_1'	K_2'	$v(\mathbf{K},\mathbf{C})$	$v_{lc}(\mathbf{K}')$	ϵ	EL	EL_{lc}	ϵ
5	3	2	4	2	3.91	1.46	.6425	.6453	−0.4	3.74	4.35	−16.3
10	6	4	8	4	7.77	3.15	.6711	.6760	−0.7	7.39	8.67	−17.3
20	12	8	15	10	14.29	6.78	.7140	.7165	−0.4	14.59	16.50	−13.1
30	18	12	15	9	15.0	9.0	.7293	.7323	−0.4	17.99	18.94	− 5.3
20	20	20	10	10	10.0	7.95	.7214	.7298	−1.2	13.13	15.78	−20.2
30	30	30	20	10	20.0	10.0	.7483	.7608	−1.7	21.23	26.44	−24.5

Table 5.1.8. Approximations of a four–machine line with additional local buffer restrictions. ν=(1.05,1,1.05,1), λ=(0.01,0.01,0.01,0.01) μ=(0.07,0.07,0.07,0.09), **eff**=(0.919,0.875,0.919, 0.9).

5.2 APPROXIMATION OF WORKLOAD–CONTROLLED LINES

A similar approximation technique as discussed in the previous section can be applied to workload–controlled lines. For workload controlled lines, too, it appears to be possible to find local control levels which result in about the same performance. The working of Workload Control is clarified in Fig. 5.2.1.

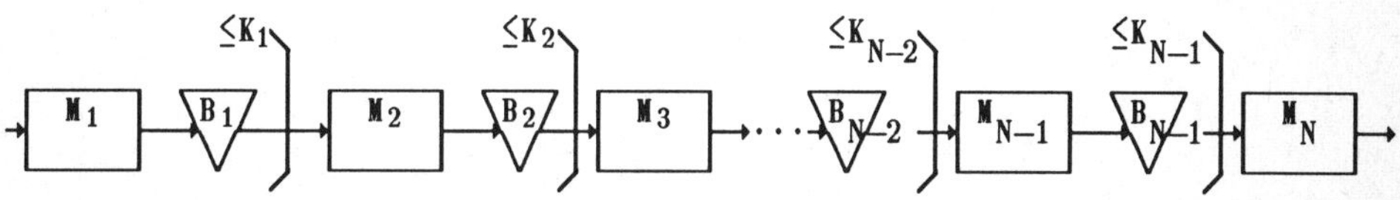

Figure 5.2.1. Flow line controlled by Workload Control

The aim of Workload Control as sketched in Fig. 5.2.1 (see also subsection 1.3.2) is to load the various machines such that the amount of work upstream of each work center is kept as constant as possible (by M_1). Machine M_1 starts producing if the content of any left echelon falls below its norm. Left echelon i is defined as the part of the line upstream of M_{i+1} (i=1,..,N–1). The inventory norm for (left) echelon i is K_i. If it is desired to control only the remaining workload of an important machine, M_j say, then the K_i can be chosen so that $K_r>K_j$, for r<j and $K_r=\infty$, for r>j.

It is again assumed that the flow is continuous and that the machines have parameters (ν,λ,μ). If left echelon i is full (at level K_i), then M_1 slows down to the rate of M_{i+1} (if necessary). It is assumed that other machines can not be blocked. Hence there are no local capacity limits. However, other machines can be slowed down by lack of input.

Note that Workload Control, with $K_1=K_2=...=K_{N-1}$, corresponds to control by BSS, with $K_1=K_2=...=K_{N-1}$. As shown in the previous section, there is no significant difference in performance of BSS and local control for the model. Even if the approximation technique used in the preceding section for determining the local buffer capacities leads to substantial relative differences, it is still possible to find good local buffer equivalents, since the errors in throughput and mean system contents have the same sign.

In order to approximate a workload–controlled line by a locally controlled one, an approximation method similar to the one in the previous section can be developed. Here, we restrict ourselves to a single example in which the local buffer capacities are constructed by trial and error. The example is given in Table 5.2.1. Results in this table are obtained for both lines by simulation with the same random numbers. 95% confidence intervals are given for the simulated throughputs.

K_1	K_2	K_3	K_4	K_1'	K_2'	K_3'	K_4'	$v(\mathbf{K})$	$v_{ap}(\mathbf{K}')$	ϵ(%)	EL	EL_{ap}	ϵ(%)
4	4	4	4	0.7	1.9	1.2	4.0	.5294(6)	.5276(5)	0.3	3.43	3.34	2.6
2	3	5	7	0.35	1.9	2.1	6.2	.5473(6)	.5438(6)	0.6	3.88	3.81	1.8
4	6	10	14	1.0	3.6	4.6	13.0	.5975(6)	.5931(6)	0.7	7.92	7.77	1.9
10	20	30	40	3.6	14.2	14.4	37.5	.6909(5)	.6849(5)	0.9	24.41	23.84	2.3
15	30	40	50	5.3	20.8	21.2	45.0	.7142(4)	.7100(4)	0.6	33.69	33.16	1.6
∞	∞	∞	∞					.7875					

Table 5.2.1. Approximation of workload–controlled line. ν=(1.05,1,0.9,0.9,1.1), λ=(0.01,0.01,0.01,0.01,0.02), μ=(0.07,0.09,0.07,0.09,0.09), eff=(0.919,0.9,0.7875,0.81, 0.9).

5.3 DISCRETE–PRODUCT MODELS

The close similarity between local and integral control demonstrated for the model used in the previous two sections may have different possible explanations.

1. The similarity may be due to the continuous product flow. The continuous–product flow causes the delay of information on demand (the speed of the last machine), which was discussed in subsection 1.3.2, to be kept to a minimum. If all machines are up, there is no delay at all. Only if at least one of the machines is down, there may be some delay.
2. Another possible explanation for this similarity is the stationarity of the model used. By simulating for 500,000 time units (about 4000 failures per machine), the system may assumed to be in equilibrium (as is clear, for instance, from the narrow confidence intervals in Table 5.2.1). Control systems like Workload Control are developed (among other things) to stabilize the system. Stationary behavior is assumed to show up earlier with BSS and workload control than with local control. This can be checked by simulating the system for a short time. The system will be closer to the equilibrium state by using Workload Control than by using local control.
3. A third factor that may be of influence is lot sizing. Lot sizing in subsequent stages may be of influence if, for instance, production of large lots is triggered in the locally controlled system and not in the integrally controlled system. This can lead to a substantial difference in work–in–process.

In this section the first argument is explored. We do this by comparing BSS and local control in a discrete–product flow line where delays occur because service times are random. In section 5.4 an example is given of a line which is not in equilibrium. This can be compared with the same line in equilibrium. Lot sizing is not further investigated, apart from the discrete–product model to be presented next, where all lot sizes are small and equal for all stages (to 1).

The influence of delays on performance differences between control by BSS and local control is investigated by using a discrete–product model in which service times are random. In this model of an (N+1)–stage flow line it is assumed that machine M_i ($0 \leq i \leq N$) has a service time that is exponentially distributed (parameter μ_i, for $1 \leq i \leq N$). The service rate of M_0 is λ. In front of M_0 there is a buffer B which is never empty. See also Fig. 5.3.1. It is assumed that the blocking process is of the transfer type, discussed in section 2.2. The capacity of the i–th waiting room B_i is K_i', including the position in front of machine M_i.

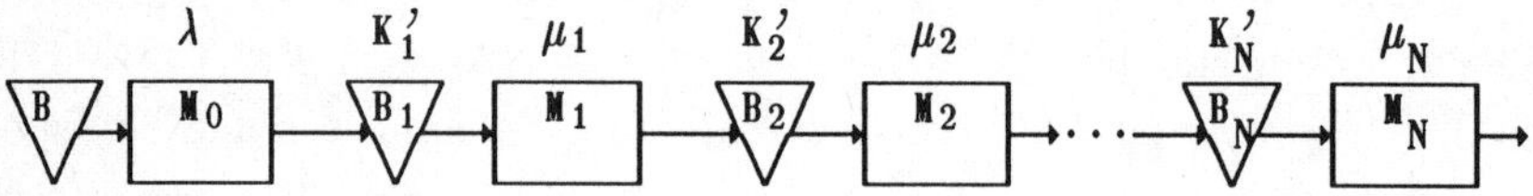

Figure 5.3.1. (N+1)–stage discrete–product flow line

The line in Fig. 5.3.1 can also be controlled by BSS straightforwardly. See Fig. 5.3.2. Machine M_i in the line in Fig. 5.3.2 is blocked (transfer blocking type) if, on service completion, right echelon i+1 is full. The capacity of echelon i is K_i, the position in front of M_i included.

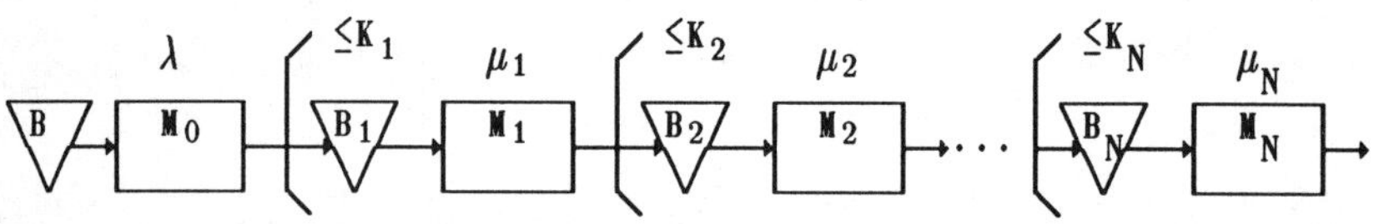

Figure 5.3.2. (N+1)–stage discrete product flow line controlled by BSS

The aim is now to find for each line as in Fig. 5.3.2, with integral restrictions K_i, an equivalent locally controlled line as in Fig. 5.3.1, with local restrictions K_i'. These K_i' are determined using a method similar to that described in section 5.1, combined with the approximation method for discrete–product flow lines with finite buffers of Perros and Snyder (1986) and trial and error.
Some results are listed for a four–stage line in Table 5.3.1. The original line is controlled by BSS, the approximating line by local control. In the approximating line the buffer capacity of the last buffer is chosen equal to the echelon capacity of the last echelon. Both lines are simulated for 100,000 units of time with the systems initially empty. EL denotes the average number of products in machine M_0 or downstream of M_0 in the line. Note that the number of products *in* M_0 is 1 at all instants for both lines.

K_1	K_2	K_3	K_1'	K_2'	v(K)	v_{lc}(K')	ϵ(%)	EL	EL_{lc}	ϵ(%)
7	5	3	3	3	.7237	.7194	0.6	6.74	6.56	2.3
10	6	4	5	4	.7739	.7779	–0.5	9.02	9.10	–0.9
15	10	5	7	7	.8320	.8306	0.2	13.01	12.92	0.7
20	15	10	9	9	.8857	.8838	0.2	16.04	16.19	–0.9

Table 5.3.1. $\lambda = 1$, $\mu = (1.1,1.05,1)$

From Table 5.3.1. it can be seen that both the Base–Stock–controlled line and the locally–controlled line again have approximately the same performance. However, for smaller echelon buffers it is more difficult to find a locally–controlled line, since we have the restriction that buffer capacities need to be integers. This is also seen from the first case listed in Table 5.3.2. The results listed in this table are obtained by one simulation run of 100,000 units of time, with the same initial state for both lines.

K_1	K_2	K_3	K_1'	K_2'	$v(\mathbf{K})$	$v_{lc}(\mathbf{K}')$	$\epsilon(\%)$	EL	EL_{lc}	$\epsilon(\%)$
3	3	2	1	2	.5847	.6109	–4.5	3.55	3.50	1.4
5	5	3	2	3	.6930	.7028	–1.4	5.05	4.93	2.4
6	6	4	3	3	.7303	.7345	–0.6	5.77	5.95	–3.1
9	7	5	5	4	.7994	.8040	–0.6	7.94	8.14	–2.5
19	16	11	11	11	.9023	.9038	–0.2	14.43	14.41	0.1

Table 5.3.2. $\lambda = 1$, $\boldsymbol{\mu} = (1,1.2,1.1)$

The approximation for the first case in Table 5.3.2 shows a relevant difference of –4.5% in throughput, which indicates indeed that for small echelon capacities the approximation by a locally controlled line becomes less accurate.

5.4 NONSTATIONARY BEHAVIOR AND CONCLUSIONS

In this section some attention is given to the possibility that integral control performs better than finite buffer control for nonstationary lines. In Table 5.4.1, the first line indicated in Table 5.1.1 is simulated for two different values for the K_i's, 10 times 2000 time units per run. The quantity σ_1^2 in Table 5.4.1 denotes the variance in throughput, σ_2^2 denotes the variance in the average line contents.

K_1	K_2	$v(\mathbf{K})$	$v_{ap}(\mathbf{K}')$	$\sigma_1^2(\times 10^{-5})$	$\sigma_{1,ap}^2$	EL	EL_{ap}	$\sigma_2^2(\times 10^{-2})$	$\sigma_{2,ap}^2$
5	2	.6916	.6919	7.5	7.6	4.15	4.17	0.40	0.70
30	12	.7571	.7585	8.1	8.6	23.42	24.00	42.1	55.3

Table 5.4.1. Comparison of the nonstationary behavior of the first line in Table 5.1.1

Results shown in Table 5.4.1 indicate that, in spite of the short run time, the performance is again about equal for both ways of controlling the line (Base–Stock and local control). This means that both lines are equally far away from the equilibrium state. The equilibrium throughput and average work–in–process are indicated in Table 5.1.2.
It is also interesting to compare the variances in throughput and work–in–process for both ways of controlling the line. Note that in Table 5.1.2 the confidence intervals for the throughput and average work–in–process are (slightly) more narrow for Base–Stock Control than for local control. In Table 5.1.4, the variances in throughput and average line contents are greater for the locally controlled line, than for the base–stock–controlled line, but the differences are again slight.

In this chapter it has been shown, that it is possible to approximate lines controlled by integral control systems like BSS and WC, by lines consisting of the same machines but controlled by finite buffers, such that the performance of the two lines is about the same. The performance measures considered are throughput and average work–in–process. This approximate "equivalence" of local control and control by BSS and WC has been demonstrated for the continuous–flow model by long–run simulations. The continuous modeling of the product flow does not have large influence on the quality of the approximation, as demonstrated in section 5.3. Also the runlength does not have much impact.

What may have some influence on the quality of the approximation is lot–sizing. If succeeding stages work with different lots have completely different processing times, then differences in performance may become greater. However, this is not investigated further.
In the simulation runs of the continuous–product lines, it was observed that the variance in average work–in–process in the workload–controlled line was always less than the variance in average work–in–process of the approximating line. However, for the BSS examples this was not *always* the case. Although BSS, WC and local control yield approximately the same throughput and average work–in–process, and hence average throughput time, it may still be so that BSS and WC in general have a positive effect on the variance of the throughput time. This variance–reducing effect is indicated by Table 5.4.1.

CHAPTER 6

CAPACITY MODELING OF MULTIPRODUCT LINES

In the previous chapters, tools have been developed for obtaining insight into the relations between system performance (throughput, average work–in–process, throughput time) and control rules. Furthermore, in Chapter 5, it was shown that as far as the above–mentioned performance criteria are concerned, for flow lines there is hardly any difference between local control rules and integral control rules.

However, only single–product machine lines were considered in the earlier chapters. Important initial objectives were to model multiproduct systems in an aggregate way, to model more general stages instead of simple machines and, in addition, to model aggregate control rules. Therefore the applicability of continuous–flow models is investigated in the present chapter for such general multiproduct lines. In Chapters 4 and 5 the considered machines are of simple up–down type. It will be argued that the approximation method developed in Chapter 4 and the equivalence of local and integral control, demonstrated in Chapter 5, also hold for general multiproduct lines. This is done by considering the differences in (control of) single product machine lines and general multiproduct lines. Some of these differences have been discussed in previous chapters. For example, in general multiproduct lines the stages are not of the extreme up–down type as in Chapters 4 and 5. However, in section 2.7 it has been demonstrated that there is a high degree of insensitivity in line performance with respect to the modeling of the speed variations in the stages. Because of this insensitivity in line performance, results of Chapters 4 and 5 also hold for stages with many speeds (modeled as in section 2.4). The approximation techniques of Chapter 4 will work for many speed stages, and the equivalence of local and integral control will also hold for such systems. See also De Koster (1986b).

There are also differences which may make it more difficult to use the model of the previous chapters and which therefore may have more impact on the results of Chapters 4 and 5:

- The single product is the aggregate of multiple products.
- In multiproduct lines there is generally some lot–sizing.
- In general multiproduct lines the stages may also contain work–in–process, while in the special one–product machine lines the stages are empty.
- The "machine" behavior is not generally stationary (this holds especially for the last machine, which represents demand).

In the two following sections attention will be given to the first two issues. They are investigated on hand of two–stage systems mainly. In sections 6.3 and 6.4 some attention is given to the other two issues. Conclusions are drawn in section 6.5. Most of the issues play a role in the differential–gear factory described in section 1.1. In this case more than two production stages are involved. It will be treated in the next chapter.

6.1 AGGREGATES INSTEAD OF ITEMS

In the previous chapters, it was assumed that a single product type is produced by the production systems. One exception is in the case of the insertion department, described in section 2.1. In the insertion department many different PCB types are produced. Aggregate modeling in this case and modeling the applied control (workload control applied on the total system contents) proved to be simple. The main reason for this is that the choice of an aggregate product unit is easy. In general multiproduct two–stage lines, the choice of an aggregate product unit is not always that easy. This choice is discussed in the present section.

In the case of a one–product situation, the production speed can be measured in units of product/hour and the buffer content in units of product. So there is no problem in this case.

Consider two subsequent stages (I and II) with intermediate buffer and several types of products flowing through the system. As in the case of one product, the buffer content can be expressed in hours of capacity of stage I as well as in hours of the capacity of stage II. The difficulty is that there is no fixed ratio of these two measures, if the ratios of the capacity requirement for both stages are not the same for all products.

As an illustration, consider the three situations sketched in Table 6.1.1. In all situations there are two stages I and II and two products, A and B.

	situation a		situation b		situation c		
	I	II	I	II	I	II	
A	1	1.2	1	1.2	1	1	(hours/unit)
B	1	1.2	0.5	0.6	1	1.5	(hours/unit)

Table 6.1.1. Capacity requirement patterns of A and B for three situations

In situation a, the products A and B have the same capacity–requirement pattern. This situation is equivalent to the one–product situation.

In situation b, products A and B have the same capacity–requirement *ratio* at capacities I and II, namely 5:6. For that reason, one hour of production of stage I is also one hour of production of stage II, independent of the product mix. This situation is also equivalent to the one–product case. The units of product have to be normalized: one unit of product has to be defined as the quantity that requires one hour at stage I (or stage II). This again leads to the capacity–requirement pattern of situation a.

Note that case of the insertion department in section 2.1 is similar to situation b. The ratio of the cycle times at horizontal and vertical inserters is the same for all PCBs (assuming that the ratio of horizontal and vertical components is 2:1 for all PCBs). In the case of the insertion department the product unit is expressed in "norm PCBs", which reduces the situation to one similar to situation a in Table 6.1.1.

Situation c is the most complex. The capacity requirement ratios differ for products A and B. To be able to aggregate in this case, one has to work with an average mix of A and B. Assume the average mix A:B is 3:2. Then the average capacity–requirement pattern is 1–1.2. This generates a certain undeterminedness at aggregate level. A buffer content of 5 (aggregate) units is in the average 5 hours of work for stage I and 6 hours of work for stage II. But if the buffer contains only product A, then it is 5 hours of work for stage II and, if the buffer contains only B, then it is 7.5 hours. The deviation of the aggregate buffer contents may deviate more from the real contents, if the capacity–requirement ratios of the products differ more and if the batch sizes become larger (since then the production mix is less stable).

A possibility for the aggregate modeling of this situation is, by measuring the buffer content in "units" and by letting the stages switch over randomly between the two speeds, so that the run time at such a speed corresponds to the average batch size of the particular product. Furthermore, the ratio of the fraction of products produced at speed 1 (corresponding to type A) and speed 2 (corresponding to type B) has to be the ratio 3:2. Such aggregate modeling was already briefly discussed in section 2.4.2. We will work it out in more detail for the case that a type A–lot consists of 6 items, a type B–lot consists of 4 products. Each of the stages switches over after production of such a lot. A switch–over takes 1 hour on average at both stages I and II. Assume for simplicity that the time to produce a lot consisting of both products A and B is exponentially distributed at both stages I and II. Also switch–over times are assumed to be exponentially distributed. The stages can be modeled as in section 2.4, with three states per stage, as indicated in Fig 6.1.1.

	stage I			stage II		
state	speed (prod/hr)	rate (hr^{-1})	trans. matrix	speed (prod/hr)	rate (hr^{-1})	trans. matrix
1	1	1/6	$\begin{bmatrix} 0 & 0 & 1 \\ 0 & 0 & 1 \\ p_1 & p_2 & 0 \end{bmatrix}$	1	1/6	$\begin{bmatrix} 0 & 0 & 1 \\ 0 & 0 & 1 \\ q_1 & q_2 & 0 \end{bmatrix}$
2	1	1/4		2/3	1/6	
3	0	1		0	1	

Figure 6.1.1. Modeling of the stages with the model of section 2.4

State 1 in Fig. 6.1.1 corresponds to production of type A, state 2 corresponds to production of type B. The speeds of the stages in these states are directly obtained from Table 6.1.1. The sojourn time in state 1 of stage I, for example, is the time needed to produce a type A–lot, namely 6 hours on average. Hence the rate in this state is 1/6. The rates in the other states are determined similarly. Determination of the transition matrix is also straightforward, apart from the probabilities p_i and q_i (i=1,2). These deserve an explanation. Let π_i be the equilibrium probability for state i of stage I (i=1,2,3). The probability p_1 ($p_2 := 1-p_1$) in Fig 6.1.1, is so chosen that the ratio $\nu_1\pi_1$ over $\nu_2\pi_2$ equals 3:2. This means that, in the long run, the ratio of products of type A over products of type B is 3:2. The equilibrium probabilities and p_1 can then be derived from the system (6.1).

$$
\begin{aligned}
\pi_1\lambda_1 &= p_1\pi_3\lambda_3 \\
\pi_2\lambda_2 &= (1-p_1)\pi_3\lambda_3 \\
\pi_3 &= 1-\pi_1-\pi_2 \\
2\nu_1\pi_1 &= 3\nu_2\pi_2
\end{aligned}
\tag{6.1}
$$

System (6.1) yields $\boldsymbol{\pi}=(1/2,1/3,1/6)^T$and $p_1=p_2=0.5$ for stage I. The probability q_1 ($q_2:=1-q_1$) in stage II can be determined in the same way. This yields $\boldsymbol{\pi}=(3/7,3/7,1/7)^T$ and $q_1=.5$. Note that, although the machine states correspond to production of either product A, B or switching over, the product types in the buffer cannot be distinguished into types A and B by stage II. It will be clear that the assumption of exponential distributions for the time to produce a lot, is not essential. It is straightforward to use arbitrary phase–type distributions. The number of machine states then increases. This way of modeling multiproduct lines is used in the differential–gear case, introduced in section 1.1 and treated in section 7.2.

The product mix in the model is based on a time average. Influences exerted by the *actual* mix on the speed pattern of the stages, and possibilities of changing the actual mix to improve the speed pattern are not taken into account. This may lead to a modeling error. The error is not serious if the production sequence is independent of the actual state of the line (buffer content, average production speeds). In such a case the throughput and average work–in–process can be properly analyzed on the basis of aggregate models alone. This is the case, for instance, if the production sequence is determined by an external random process, like customer orders.
However, in reality, the production sequence can be adapted to realize better machine utilizations. In general, the assumption of a randomly determined production sequence, based on an average mix, leads to a useful lower bound for the performance. The lower bound will be close if the capacity requirement ratios do not differ much.

In the multistage case the same approach can be used. See section 7.2.

The situation becomes even more complicated if the stages consist of more than one capacity, for instance men and machines or different types of machines. A way to model simultaneous requirement of men and machines, is by two "machines" in series, without intermediate buffer. If there is shortage of either man or machine capacity, the entire stage will slow down or stop completely. The possibility of aggregating such different capacities to one capacity, depends also on the differences in capacity–requirement ratios over the products and on the stability of the production mix.

6.2 LOT–SIZING

Production of different products with the same capacities often leads to production in lots. After production of a lot of a certain type, a capacity is switched over. In the insertion–department case in section 2.1, production was in lots. Modeling this department educed no problem, because the downstream stage (vertical insertion) could keep on producing if less than a complete *production lot* was available in the buffer. But an adaptation in the model had to made for the *transportation lot*, the unit that arrived in the physical buffer. This unit was a single PCB. The adaptation was to choose the buffer capacity in the continuous–flow model lower than in the real system. It was argued that the real department had about the same performance as the continuous–flow model with a decreased buffer upper bound. It is equivalent to raise the buffer lower bound in the model instead of decreasing the buffer upper bound. In subsection 2.5.1, it is argued that two equal two–stage lines, but with buffer ranges [a,b] and [C+a,C+b] ($a<b$, $C \in \mathbb{R}_+$), respectively, and the same slowing–down behavior at lower and upper bounds of the buffer are *buffer–shifted–equivalent.* This means that they have the same throughput and furthermore that the mean buffer contents are shifted over C.

In this section it will be demonstrated that continuous modeling of two–stage situations with lot–sizing leads to models with adapted buffer bounds, namely increased buffer lower bounds. The buffer lower bound in the model equals the average buffer stock, solely due to the lot–sizing effects. This adaptation of the buffer lower bound is based on the idea that buffer stock caused by lot–sizing can scarcely be used for buffering against disturbances in capacities, such as machine failures. The lot sizes may be different for different products and there may also be lot–size differences between stages.

For simplicity we will restrict ourselves to two stages only. In the differential–gear case we have to do with lot–sizing in more than two stages. The ideas developed here, are applied in that case. The following aspects can be distinguished:

1. Lot–sizing in only one stage, or lot–sizing in both stages
2. Production lots only, or production and transportation lots
3. In the case of lot–sizing in both stages, are the lot sizes coordinated in quantity or not? Can batches for the same product be started (nearly) simultaneously at subsequent stages (time synchronization)?

The more coordination there is between the production lots of the different products in the two stages (in quantity and starting time), the less the increase of the lower buffer bound in the continuous model has to be. If besides production lots there are also transportation lots (that is, the products arrive in the physical buffer in discrete quantities), then the lower buffer bound has to be increased even more. The following situations are considered:

a) coordinated lots in both stages: $Q_{I,i}=Q_{II,i}$, where $Q_{j,i}$ is the (production) lot–size of product i at stage j (j=I,II);
b) $Q_{II,i}=0$, no lot–sizing at stage II;
c) coordinated lots in both stages: $Q_{I,i}=n_iQ_{II,i}$, with n_i integer;
d) uncoordinated lots

Situations a and b are discussed for the cases with and without transportation batches. The transportation lot–size is assumed to be independent of the product type. It is denoted with Q_T. For these two situations it is shown, that the case with transportation lots is buffer–shifted equivalent to the case without transportation lots, with an extra increase of the lower buffer bound. A related result is derived by Hontelez (1985). He considers a single–product two–stage line with transportation lots. In this line, the first stage is perfectly reliable and the second stage is unreliable, with exponentially distributed life and repair times. Hontelez shows that this situation can be represented by a continuous model without transportation lots, but with a raised buffer lower bound (or actually a lowered buffer upper bound).

No attempt is made to evaluate the adaptation of the buffer lower bound in the model precisely for all possible cases. Although the average stock due to lot–sizing effects may be hard to evaluate in particular situations, it is shown that at least upper bounds can be obtained relatively easy. An upper bound for the stock due to lot–sizing leads to a lower bounds for the throughput. Since we are only interested in the stock in the system due to lot–sizing, it is assumed that there are no machine failures (or switch–over delays, which can be modeled as a special type of failure). The influence of machine failures on the quality of the model is discussed at the end of this section. Another assumption in this section is that capacity requirement ratios for different capacities are more or less equal for all products. The results are based on De Koster and Wijngaard (1988a).

a. $Q_{I,i}=Q_{II,i}$, *for all i.*

Consider a two-stage continuous-flow system with lot-sizing in both stages, without transportation lots. Hence, the products arrive in the buffer in infinitesimally small amounts and are taken out of the buffer in infinitesimally small amounts. The lots at the two stages are the same for all products. Assume furthermore that a lot partly finished by stage I can be started processing by stage II. In that case the distance of the two stages from each other (measured in stored capacity) can be very small, just as in the case without lot sizes: the lots are produced (almost) simultaneously by both stages. So, if a continuous model is used, the minimum buffer level has to be set equal to 0 in this case.

In case products arrive in transportation lots of size Q_T in the buffer, the capacity distance of the stages becomes larger. For all product types, stage I remains somewhat ahead of stage II. This amount can be reasonably estimated by Q_T. Hence, in a continuous-flow model the lower buffer bound has to be set equal to Q_T.

b. $Q_{II,i}=0$, *for all i.*

Consider the case where the second stage has constant production rates for all products. The first stage produces in batches. It is assumed that a partly finished run can be counted as being part of the buffer content. This means that as soon as one product of that batch is finished it is available for consumption by stage II. Hence there are no transportation lots. Stage II consumes products from the buffer cyclically: ABCABC.. (arising from an external demand, for instance), without switch-over delays. Stage I has to fulfill the needs of stage II. The decrease in buffer content is the result of a sequence of many (infinitely) short production runs of that product. Let p_i be the speed of stage I for product i and let d_i be the speed at which product i is used by stage II. The inventory pattern for the case of three products is then as sketched in Fig. 6.2.1. The inventory pattern for a single product is sketched in Fig. 6.2.2.

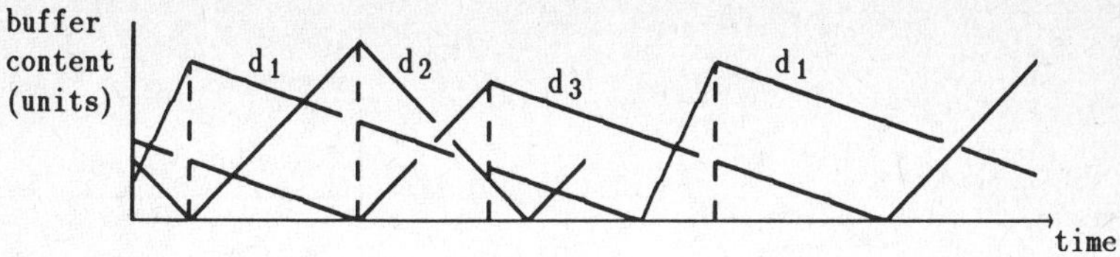

Figure 6.2.1 Inventory pattern in the case of constant consumption by stage II and no idle periods of stage II

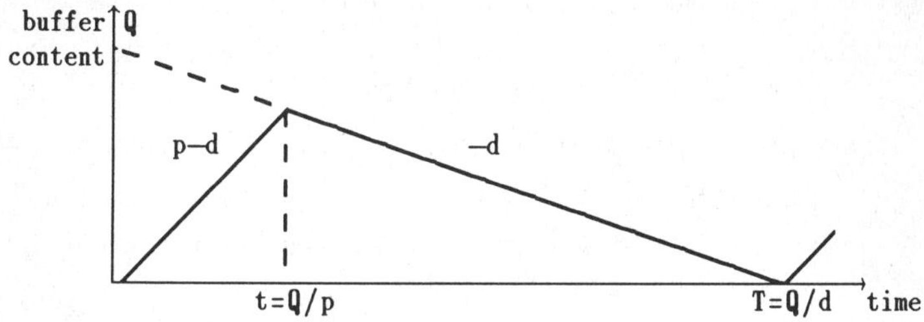

Figure 6.2.2. Single product, constant consumption rate

In Fig. 6.2.2, T is the cycle time and t is the run time for the product. Stage I does not produce the product during T–t. The average inventory is equal to

$$\frac{Q}{2}(T-t)/T = \frac{Q}{2}(p-d)/p$$

Hence, in the multiproduct situation the average inventory is equal to

$$\sum_i \frac{Q_i}{2}(p_i-d_i)/p_i \tag{6.1}$$

A straightforward choice for the buffer lower bound in a continuous flow model of this situation would be the amount in formula (6.1). The difference with the real system is, that it may be possible that stage II is starved when the buffer is empty, while in the model stage II is starved when the buffer level equals (6.1). Therefore, the choice of (6.1) as buffer lower bound in the continuous flow model only results in a good representation of the system, if there is little variation in the amount of stock due to lot-sizing: if this amount were constant (and equal to (6.1)), the modeling would be exact.

In order to investigate the variation in the stock due to lot-sizing, consider the case of two products and no idle periods, sketched in Fig. 6.2.3a.

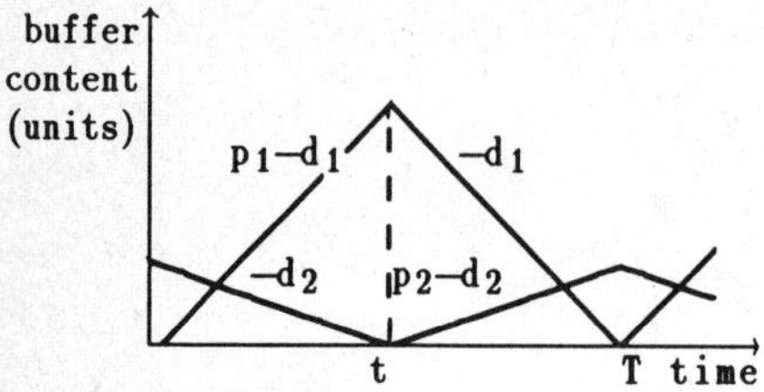

Figure 6.2.3a.

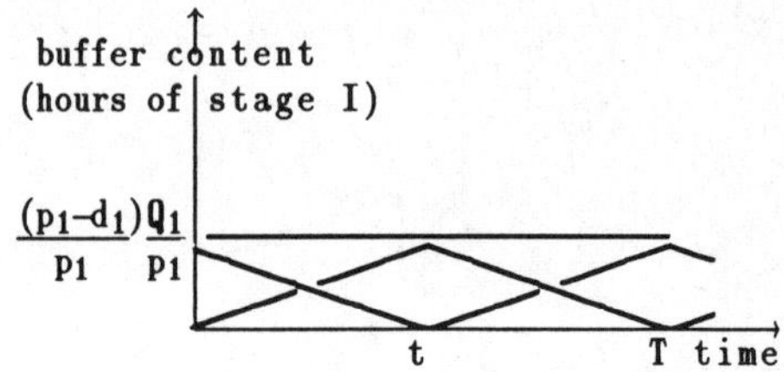

Figure 6.2.3b

Two–product situation, without idle time of stage I. The cycle time $T=Q_1/d_1=Q_2/d_2$, $t=Q_1/p_1=Q_2/d_2-Q_2/p_2$ is the run time of product 1, T–t is the run time of product 2.

The average inventory for the situation sketched in Fig. 6.2.3a (measured in units) is given by formula (6.1). Measured in units, the actual inventory due to lot–sizing varies between $Q_1(p_1-d_1)/p_1$ (all products of type 1) and $Q_2(p_2-d_2)/p_2$ (all products of type 2). If the number of items increases, then the fluctuations in buffer content will decrease and the actual inventory will be close to (6.1). Measured in hours of production of stage I, the actual inventory is constant, even in the case of only two products. This is sketched in Fig. 6.2.3b. Note that it follows from the full utilization of the stations that $Q_1/d_1=Q_2/d_2$ and $Q_1/p_1=Q_2/d_2-Q_2/p_2$. This implies that $\frac{p_1-d_1}{p_1}.\frac{Q_1}{p_1} = \frac{p_2-d_2}{p_2}.\frac{Q_2}{p_2}$, and therefore the stock is constant, measured in hours of production of stage I.

Hence, it follows that if the stages have high utilization rates (relatively small idle periods), then the stock due to lot–sizing will be nearly constant (measured in hours of stage I) and hence the choice of (6.1) as buffer lower bound (6.1) in a continuous–flow model will yield a good approximation.

Assume now that the products arrive in transportation lots of size Q_T in the buffer. The transportation lot size is the same for all products. In this case the buffer lower bound has to be set somewhat larger as indicated in (6.1), just as in the case of equal lot sizes in the stages. An increase of (6.1) by Q_T will be reasonable in most cases.

c. $Q_{I,i}=n_i Q_{II,i}$, *for all* i, n_i *integer.*

If the lot–sizes in stage I are n_i times the lot–sizes in stage II, then there is again a certain coordination: the n_i lots in stage II form precisely one lot of stage I. First, it is assumed that time coordination is also possible. The case in which there is no time coordination is discussed later. The resulting inventory pattern for one product, with $n_i=4$ is sketched in Fig. 6.2.4, for the case of a regular consumption by stage II of this product. The production speeds are assumed to be equal for both stages. Stage II already starts production as soon as the first (infinitely small) unit of product is finished by stage I, and hence there are no transportation lots. It is furthermore assumed that the second stage consumes lots in FCFS (First Come First Served) order: lot A–lot B–lot C–lot A–lot B–lot C–.. .

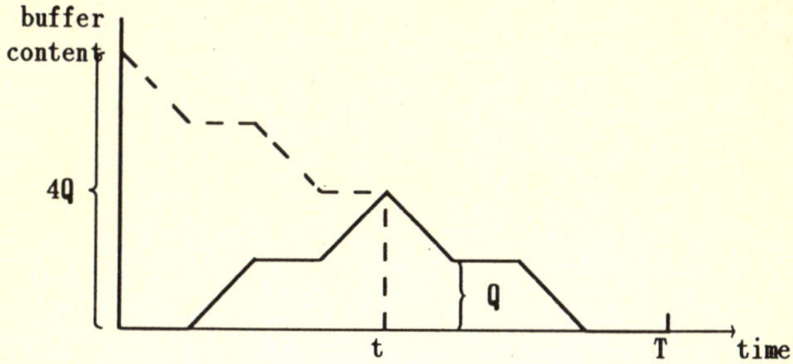

Figure 6.2.4 Ideal inventory pattern for one product, in case $Q_I=nQ_{II}$ and synchronization

In the situation sketched in Fig. 6.2.4, with alternating runs for the product and other products, the average inventory for one product is equal to (indices omitted)

$$\frac{Q_{II}}{2}(n-1)\frac{T-t}{T} = \frac{Q_{II}}{2}(n-1)\frac{p-d}{p},$$

where p is the production rate of stage I and d the average–use rate of the product. T is the cycle time and t is the run time of the product. Note that, in the limiting situation that the second stage becomes faster, or that the batch sizes at stage II become smaller, the situation sketched in Fig. 6.2.4 tends to the situation sketched in Fig. 6.2.2. In case of many products, where stage II consumes batches of different products in a cyclic order, a reasonable choice for the lower buffer bound, will be

$$\sum_i \frac{Q_{II,i}}{2}(n_i-1)\frac{p_i-d_i}{p_i}.$$

Note that this formula corresponds to the lower buffer bound in situation *a* in case $n_i=1$, for all i.

d. Uncoordinated lots.

The situation of uncoordinated lot sizes (in quantity) in the stages is more difficult than the previous ones. It is more difficult to derive good buffer lower bounds for a continuous model. The average stock in the system due to lot–sizing effect will be larger than in the case of coordinated lots. This leads to a higher buffer lower bound in a continuous–flow model. It will depend on whether there is time synchronization or not: is it possible to start a batch in stage II nearly simultaneously with the same batch in stage I, or not? In the case of time synchronization, the buffer lower bound will be lower than it would be without time synchronization. Furthermore, the stock due to lot–sizing will be less constant, which means that the choice of a hard buffer lower bound leads to an underestimation of the throughput. For a more extensive treatment of this case, see De Koster and Wijngaard (1988a).

The situations with coordinated lot sizes sketched above show that it is indeed possible to use continuous models in the case of lot–sizing and capacity disturbances, by adjusting the buffer lower bounds. In such a continuous–flow model, the lower bound on the buffer content has to be chosen equal to the average stock due to lot–sizing effects alone. This average stock has to be determined by assuming that no capacity disturbances occur. Generally, the stock caused by lot–sizing effects can only be used to a limited extent to buffer against capacity disturbances (see Fig. 6.2.3b).
Although it is not always possible to determine this lower bound precisely, it is fairly straightforward to overestimate it, so that at least a *lower bound* for the throughput can be derived.

6.3 WORK–IN–PROCESS

Production stages in general multiproduct flow lines may consist of several machines and buffers and are not empty therefore, but have work–in–process generally (see Fig. 6.3.1).

Assume first that each production stage can be characterized by one (bottleneck) capacity. Hence the other capacities in the stage are much faster for all products. The work–in–process in the stage consists of two components. In the first place, the work that has been executed already by this capacity, but has not yet been transferred to the buffer (EW) and in the second place the work that has not been executed yet, that is the remaining work (RW).

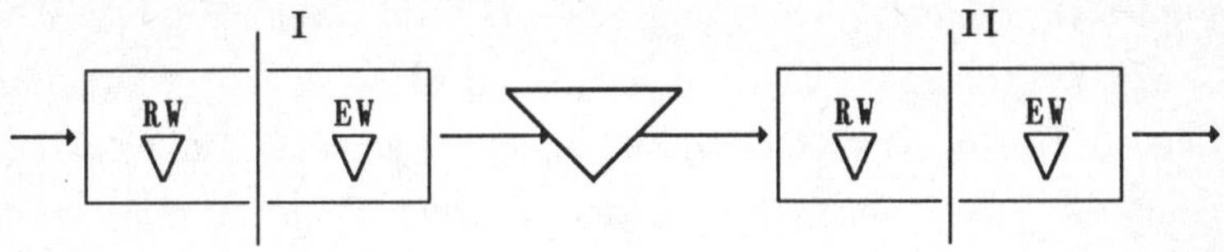

Figure 6.3.1 Production stages with work–in–process

A normal procedure in such a situation is to apply workload control (see sub–section 1.3.2), in order to keep the remaining work constant. The effect of workload control is that not only the remaining work, but also the executed work remain fairly constant (at their norms RWN and EWN). If the remaining work and executed work are kept strictly constant, then, in a continuous–flow model, the capacities can be positioned at the beginning of the production stages, instead of somewhere in the middle (see Fig. 6.3.2).

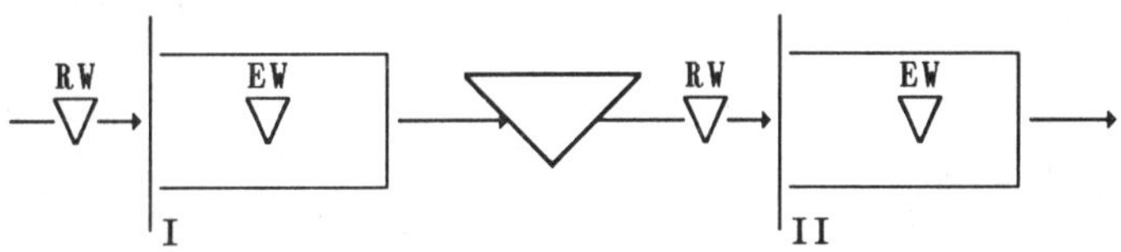

Figure 6.3.2 Workload–controlled stages

The distance between capacities I and II, measured in stored capacity, is then equal to RW(II)+EW(I)+ buffer content. The system in Fig. 6.3.2 can be modeled as a system without work–in–process, but with the lower bound for the buffer content equal to RWN(II) + EWN(I). In the case of a less strict workload control, where the remaining work is only kept *below* a certain level, it is probably better to model the system in Fig. 6.3.1 in the following way as a system without work–in–process, with four "machines" instead of two (see Fig. 6.3.3).

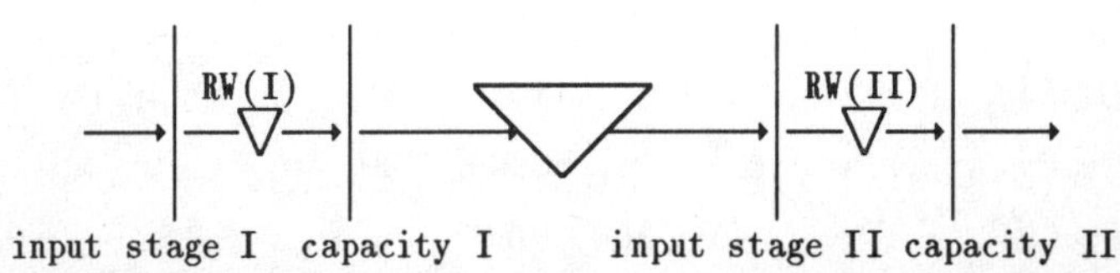

Figure 6.3.3 Split stages

The buffer between capacity I and input stage II contains EW(I) and the inventory between the two stages. The inventory between input stage i and capacity i contains RW(i), i=I,II. The input machines have to be modeled as high-speed ones and are completely reliable. The input of stage i becomes blocked if the line content between input stage i and capacity i equals RWN(i). The lower bound of the buffer content between capacity I and input stage II has to be adjusted for the average amount of work executed by stage I, since stage II may become starved while there is executed work in stage I.

6.4 NONSTATIONARITY

One difficulty in using the results developed for the continuous–flow model in aggregate production control of logistic lines is that in many cases demand and capacity in aggregate production control are dynamic, while the models used in the previous chapters are stationary. This difficulty arises only, of course, if the *aggregate* behavior is nonstationary. It should be realized that aggregate demand behavior is often more stable than the demand behavior of individual items. For instance, some products may be at the end of their life cycles, but other products at the beginning of theirs, so that aggregate demand, expressed in volumes, may be stationary. The remaining problem is how to check that the demand volume still has the same distribution so that the model parameters are still valid.

Nonstationarity may yield a difficulty in application of the model. This difficulty has of course nothing to do with the continuous character of the model, it would also be there in the case of discrete modeling (queueing–type models). Whether the conclusion of Chapter 5, namely that local information is sufficient in the case of stationary aggregate production control, can be extended to the dynamic case is still an unanswered question. It depends of course on the degree of dynamics. It may be possible to apply the tools developed in Chapters 4 and 5 quasi–statically. That is, to apply the approximation methods separately for subsequent system parameters. The problem is that in those chapters only stationary behavior of the models is studied. Nevertheless, the following simulations make clear that stationary behavior occurs relatively fast in these models. Reasonable stationarity already appears after about 5000–20000 time units, as is demonstrated by the simulations of two two–stage lines, listed in Table 6.4.1. For these particular lines this corresponds to about 50–200 failures per station. With reasonable stationarity we mean that the differences between throughput and equilibrium throughput and between mean buffer contents and equilibrium mean buffer contents are "slight".

		exact	simulation time (units) 1000	5000	10000	20000
case I	v	0.8061	0.8286	0.8427	0.8374	0.8132
	EX	4.160	2.363	4.425	4.159	4.219
case II	v	0.8702	0.8919	0.8998	0.8786	0.8761
	EX	1.735	1.133	1.654	1.749	1.754

Table 6.4.1 Simulation of two two–stage lines for different simulation times. Case I : $\nu_1=1$, $\nu_2=1.1$, $\lambda_1=0.01$, $\lambda_2=0.02$, $\mu_1=\mu_2=0.09$, K=10. Case II: $\nu_1=1$, $\nu_2=1.1$, $\lambda_1=0.01$, $\lambda_2=0.007$, $\mu_1=\mu_2=0.09$, K=10. For notations see section 2.3.

In order to apply the models dynamically, the term at which model stationarity occurs has to be shorter than the term at which system parameters change.

There may also be possibilities of making the demand stationary by applying a certain smoothing, or by making the whole system more or less stationary by changing the capacities simultaneously with the demand, so that the utilization rates remain constant. With respect to the first approach, stationarity may be obtained by considering the system over a longer time horizon and production–to–stock. Production may then be on the basis of a stationary program. Think also of the Kanban approach, where the production volume is more or less constant and where the production mix is decided monthly.

A case where the results can be applied is that of the differential–gear factory of section 1.1. The production volume is determined by the number of trucks in the assembly plan. The truck manufacturer decides each year the number of products to assemble each day. Such a decision makes all preceding stages stationary. The case will be described and analyzed in Chapter 7.

6.5 CONCLUSIONS

In the previous sections it has been argued that continuous production models may be the right ones for modeling the capacity level in multiproduct production systems. That suggests that the conclusion in Chapter 5, that local control is roughly equivalent to echelon control and workload control, also holds for aggregate goods–flow control. That does not mean that only local information is needed for goods–flow control as a whole. Generally it will be necessary to coordinate the timing of the lots in the various stages (see section 6.2). And that requires information about the delivery schedule in all stages. So, at item level the required information is integral, at capacity level the required information is local.

A possibility for realizing coordinated timing is by relating the production lots of the stages to a fixed delivery schedule: each lot covers a certain part of that schedule. Assume that the first L_{II} lots of stage II cover the delivery schedule up to time T. And that there are at least L_I lots of stage I required to cover the same part of the delivery schedule. Then coordinated timing is implied if lot L_I of stage I is produced just before lot L_{II} of stage II.
This also shows that it is not essential for the lots to be constant over time so long as the lots cover a given part of the delivery schedule. In the case that lot sizes are variable, use of a continuous–flow model with an increased buffer lower bound may still be possible. The approximation results however, will be less than in the case of fixed lots, since the average buffer stock due to lot–sizing effects will be less constant.

Although it may be difficult to apply the continuous–flow model in an environment with many different products with different capacity–requirement ratios, or lot sizes between the stages, the results of sections 6.1 and 6.2 show that the model may anyway be used to obtain lower bounds for the system performance.

CHAPTER 7

CASE STUDIES

How the tools developed in this text can be used for the analysis and design of real production systems will be demonstrated in this chapter by means of two cases.

The first case, described and analyzed in section 7.1, is a design problem. The problem is to design a production system so that a prespecified performance is realized. In this production system there are certain degrees of freedom, like buffer spaces, system layout (to some extent), number and allocation of operators and the control rule by which products are released to the system. This case is based on Giesberts et al. (1987).

The second case is treated in section 7.2. It is an existing factory (the differential–gear factory, discussed in section 1.1). This factory is part of a chain of factories producing trucks. It has externally determined targets, such as a desired (aggregate) throughput ("trucks" per year), delivery performance measures like short lead times and furthermore little work–in–process. The problem here is to provide production control rules at capacity level to realize the required performance.

In both cases the production systems produce different products. In order to analyze the systems at capacity level, these products have to be aggregated. The aggregation procedure is different for these cases. In the first case aggregation difficulties are circumvented by switching over from one set of products to another *between* shifts. In the second case, there are many different products with different capacity requirement ratios. The applied aggregation procedure in this case is the one described in section 6.1.

In the previous chapter, problems of applying the single–product continuous–flow model to multiproduct production situations were discussed. Solutions to aggregation problems, problems of lot sizing and work–in–process were given. In the differential–gear factory these problems are treated in a multistage environment.

The aggregation method for the first case is validated by comparing the results with detailed simulations. For the second case only partial validation is given (only the pre–operations and teething–capacity group were simulated). The case is intended as a demonstration that modeling and analysis at capacity level (using tools of this text may be possible for a such complicated situation.

7.1 A COIN–VALIDATOR PRODUCTION LINE

7.1.1 Case description

The coin–validator production line is still in the design phase. It is going to be part of a production system which produces machines in which coin validators are used (for instance coffee machines). The coin–validator line produces a large number of different coin–validator types. The layout of the line is sketched in Fig. 7.1.1.

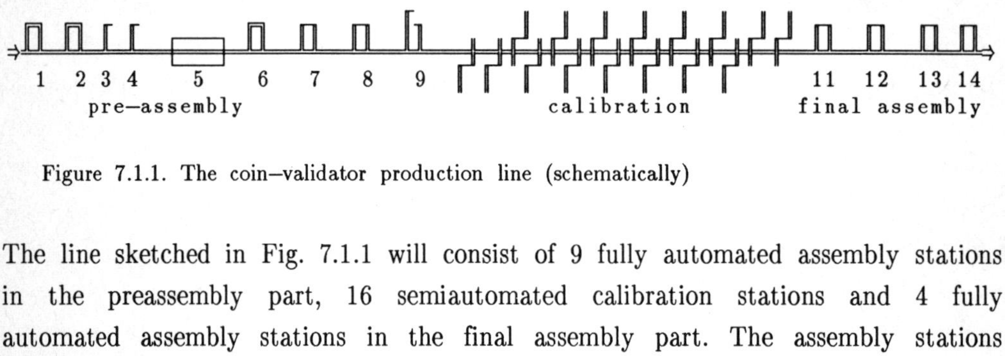

Figure 7.1.1. The coin–validator production line (schematically)

The line sketched in Fig. 7.1.1 will consist of 9 fully automated assembly stations in the preassembly part, 16 semiautomated calibration stations and 4 fully automated assembly stations in the final assembly part. The assembly stations assemble different part types on the main product, without switch–over delays. A product– type detector communicates the product type to the assembly station, which chooses the right part from the different part–type bins attached to it. Each coin validator consists of 18 parts. The total number of different parts is 28. A finished coin validator consists of two major parts: a mechanical and an electrical part. The total number of mechanically different products that can be assembled is 320. The electrical part is a so–called Programmable Read–Only Memory (PROM). This PROM is programmed, as required for the country to which the final product is sold. This is done at any of the calibration stations set up for that particular country. Part of the calibration operation is done by hand, the major part of the work is done automatically. A single operator can serve several calibration stations. Switching a certain calibration station from one country calibration over to another is time consuming (in the order of an hour) compared to the calibration cycle time (6–8 minutes). Mechanically different types can be handled by a calibration station without switch–over delays, as long as the products are of the same country type.

All stations are connected by a conveyor belt. The assembly stations are all situated directly at the line, connected in series. They have in the present design proposition a buffer capacity of 6 products each, plus a working position at the station. The calibration stations are all situated off the main line, that is, the products have a bypass possibility. In the present design phase two buffer positions per calibration station are planned, plus a working position at the station. Each product is, depending on the country variant, automatically allocated to a calibration station for that country that has room in its buffer and the smallest remaining workload. If there is no such calibration station available, the preceding assembly station (station 9) is automatically stopped. Since the products have bypass possibility at the calibrators, the calibrators actually work in parallel.

All assembly operations are carried out by robots and all assembly cycle times are deterministic. The slowest station in the preassembly part has a cycle time of 20 seconds. The slowest station in the final assembly part has a cycle time of 19 seconds. All assembly machines are designed to have a *station availability*, defined as machine up time/(machine up time+machine down time) of 95%. The cycle time of a calibration station depends on the country type and varies from 6 to 8 minutes. The cycle times of all machines are listed in Table 7.1.1. The numbers 1,..,9 refer to machines in the preassembly, number 10 refers to a calibrator and numbers 11,..,14 refer to assembly machines in the final assembly.

machine number	1	2	3	4	5	6	7	8	9	10	11	12	13	14
cycle time (sec.)	20	18	12	12	8	12	20	20	9	360–480	8	13	7	19

Table 7.1.1. Cycle times of the machines in seconds

The speed of the conveyor belt is expected to be about 12.5 cm per second.

Products are placed on product carriers, which have a length of 25 cm. Hence, the conveyor–belt cycle time is about two seconds per buffer position.
As mentioned earlier, the cycle time at a calibration station varies between 6 and 8 minutes, depending on the country type. About 30 seconds of the first part of the cycle time is taken up by a manual operation. The remaining operations are done automatically and have constant operating times. If an operator is available, there may be a small standard deviation in these 30 seconds. If there is no operator available, the calibrator has to wait till an operator arrives. The speed of the calibration part as a whole varies from 2.67 to 2 products per minute if all calibrators calibrate 6 minutes or 8 minutes, respectively (assuming that sufficient operators and products are available). Hence, it is clear that the calibration part is the bottleneck in this line.

The production in this line will take place in 3 shifts of 7.5 hours per day, 5 days per week. In view of the long set–up times of the calibrators when changing between country variants, it was decided to set–up the calibrators between two shifts only. As mentioned earlier, the coin–validator line is only one of the departments in the factory. The production in the factory as a whole is coordinated with an MRP control system. Every period (week) the departments receive an updated production plan. The production plan has a horizon of several weeks and specifies per period what products should be produced and in what amounts. Within each period the coin validator department is free to schedule the production, independently of decisions made in other departments, on condition that the production plan is realized.

7.1.2 Design problems

The major objective for the coin–validator department is to meet the production plan. At this moment, the annual mix over the different country variants is not known. Only the total annual required production volume is known, and that equals 550,000 products. Including set–up capacity losses, scrap and rework, the required annual production volume of 550,000 products leads to 122 products per hour. 10% slack for rework and scheduled maintenance of the production system is included in this number. Depending on the annual product mix (over country types), the required production rate leads to a required line efficiency varying from 102% to 76%. The line efficiency is defined as

$$\text{line efficiency} = \frac{\text{obtained throughput}}{\text{maximum throughput}}.$$

The maximum throughput is the number of products that can be produced per hour at the bottleneck, the calibration part, if it is fully occupied and there are no disturbances. Hence, it is clear that if the annual product mix consists of country types with 8 minutes calibration time in all, then the production volume can not be realized. This is however most unlikely.

Since calibrators are switched–over between shifts, it is known within a shift what product types have to be calibrated. The product–release function must ensure of sufficient supply of the right products at the assigned calibration stations. Preventing idleness of the calibration part and waiting for operators is the most important control issue for this department. The installation of large buffers at all stations would of course help to solve the problem, but it has to be realized that buffers are expensive and that the space for buffers is restricted. In view of this, the major problems are:

1. to determine whether the planned number of buffer positions at the calibration stations is sufficient;
2. to determine the number of operators;
3. to develop a control system which guarantees that the required throughput can be realized.

The product–release structure is therefore discussed in subsection 7.1.3. In subsection 7.1.4 attention is given to determination of the number of operators. In the present proposal there is storage room for three products at each calibration station (see Fig. 7.1.1), including the operating position. It will be shown that this is sufficient to attain the required performance.

7.1.3 Control structure

Before the capacity model of the coin–validator line is discussed in detail, a number of general remarks about the proposed product–release structure have to be made.

As already mentioned, the major task of the product–release function is to utilize the calibration group as well as possible. This means that, for all calibration stations, the corresponding country variant has to be set up with the proper frequency. The utilization of a calibration station can be controlled by keeping the remaining workload of this station in the production system at a constant level. Let n_i be the number of calibration stations, calibrating country variant i. The structure of the set–up rule is then

> "Set up country variant i as soon as there are less than $x \cdot n_i$ noncalibrated country variants in the production system",

where x is the remaining workload per calibration station (in numbers of coin validators). The status of a product changes from not calibrated to calibrated the moment the calibration is finished. This set–up rule stabilizes the system and sets up each individual country type frequently enough so that all calibration stations receive the right country variants in time, provided the throughput times are not too long. If the remaining workload per calibration station x is fixed at three noncalibrated products (at the most), then no blocking will occur at the pre–assembly by full calibrator buffers. Therefore, the remaining workload is set at three noncalibrated products per calibration station and hence 48 products for the calibration part as a whole.

7.1.4 Problems involving operators

The real line efficiency is affected by the availability of operators at the calibration part. In general, every product will have a certain waiting time for service, which depends on the number of operators and on the way the operators are assigned to the calibration stations. In this subsection the number of operators and their allocation over the calibration stations are determined so that the *calibration efficiency* is sufficiently high. The calibration efficiency is the fraction of time that the calibration part as a whole can operate, given that noncalibrated products are always available. If the waiting time of operators for items at calibration, and the waiting time of items for operators, given sufficient item availability, were independent, the real line efficiency would be the product of the line efficiency with 100% operator availability and the calibration efficiency. However, these waiting times are mutually dependent. If there is some waiting time for operators, the utilization of the calibration part will increase. If the calibration part is not fully loaded, the waiting time for operators will decrease.

It is now assumed that that these two kinds of waiting times are independent. An upper bound for the real line efficiency is then the minimum of the two efficiencies, and a lower bound is the product of the two efficiencies.

The calibration efficiency can be approximated using the tables in Peck and Hazelwood (1958). In using these tables, it is assumed that the operator times are 30 seconds per product. Letting the number of operators vary we obtain the following results.

Assume that four operators are assigned to all 16 calibrators. Then it follows from tables in Peck and Hazelwood that the calibration efficiency is 100.0%. If these four operators are split into two groups, with each group responsible for 8 machines on one side of the conveyor (see Fig. 7.1.1), then the calibration part efficiency varies from 99.8% for an average calibration time of 8 minutes, to 99.5% for an average calibration time of 6 minutes.

If a single operator were available for each group of 8 calibrators, then the calibration part efficiency would vary from 96.7 to 93.8%.

It is thus advisable to use four operators. Since it is not easy to allocate operators to calibration stations at different sides of the line, it is best to allocate two operators to each group of eight calibrators at one side of the line.

In view of the high calibration–part efficiency for the case of four operators and of the negative correlation between waiting time for items by operators and waiting time for operators by items, the modeling and analysis of the line in the next sub–section is based on the assumption that there is no waiting time for operators by items. The modeling is such that the performance of the real system will be better than that of the model.

7.1.5 Modeling the coin–validator line

The coin–validator line is modeled as a continuous–flow line, with machines of the kind introduced in section 2.3. The coin–validator line is so modeled, that the performance of the real system is better than that of the model.

Only the preassembly stations and the calibration part are included. Final assembly is relatively fast compared with these parts and will have no effect on the line throughput. It only influences the throughput time. The model is sketched in Fig. 7.1.2.

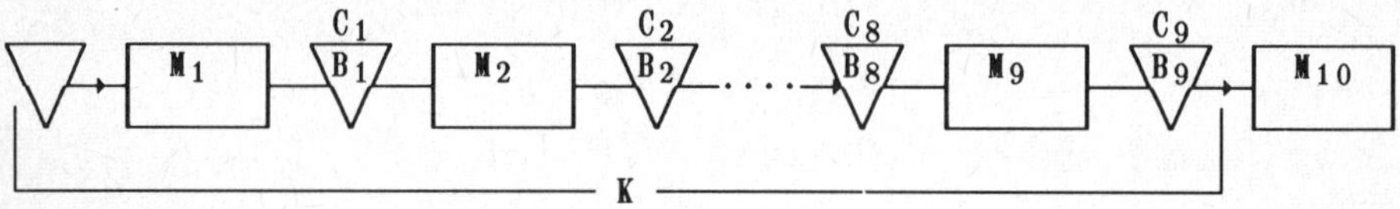

Figure 7.1.2. Model of the coin–validator line. C_i is the capacity of buffer B_i, K is the workload norm.

The machines are modeled for constant speeds, which can be obtained from Table 7.1.1. Life and repair–time distributions of the preassembly machines are not known. It is assumed that they are exponential, with parameters λ_i and μ_i, respectively. The parameters λ_i and μ_i are chosen so that the percentual machine availability $(\frac{\mu_i}{\lambda_i+\mu_i})$ is 95%, for all preassembly machines. The aggregate flow in the model is controlled by workload control. The total work–in–process in the system is kept at level K (expressed in numbers of coin validators).

The problem is to determine the parameters in the line in Fig. 7.1.2. The intermediate buffer capacities are fixed at 7 products in the preassembly. Hence $C_i=7$, for i=1,..,8. It is assumed that products finished by some station are immediately available for the next downstream station. The error made in this way is not serious. The crossing time per buffer position is about two seconds and hence the crossing time of a completely empty buffer is 14 seconds, which is small compared with the 20 seconds cycle time of the slowest machines in the pre–assembly. Furthermore the conveyor is in general not completely empty, so that the delay of assembly stations caused by the conveyor will be slight. Hence the throughput will hardly be affected by the speed of the transportation system.

Since we want to model the department at capacity level, aggregation over the products is necessary. From a control point of view all different mechanical variants are identical, since

- all different parts are assumed to be present at the line,
- there are no differences in cycle times between the different mechanical product variants and
- there is, at all stations, no switch–over time between mechanical variants.

Hence the only remaining problem is the aggregation over the country variants.

The calibration part is modeled as a single, perfectly reliable machine with the same constant speed for all country types. To determine the production rate in a certain shift of this so–modeled *calibration station* for aggregate products we proceed as follows. Let c_i be the cycle time of a calibrator calibrating country variant i in a certain shift. The production rate ν for aggregate products is determined to be the time–weighted production rate of all calibrations in the shift. Hence, $\nu = \sum_i \frac{n_i}{16} \frac{1}{c_i}$, where the summation is over the different country variants calibrated in the shift, n_i is the number of calibrators calibrating country variant i. The aggregate product is interpreted as a mixture of the different country variants, with the corresponding weight for country variant i being $\frac{n_i}{16}$. Note that the production rate is shift–dependent. The aggregation procedure differs with the methods described in Chapter 6, since here the calibrators are parallel, while in Chapter 6 it was assumed that a single station processes all product types.

The next problem is the translation of the set–up rule in the real system to a set–up rule in the model. The set–up rule in the *model* is to set up a new (aggregate) product as soon as the number of aggregate products in the system is less than K. The constant K depends on the real number of products in the system (48), but it is chosen smaller than 48 for reasons discussed below.

It is assumed in the model that, of the 48 products in the system, 16 are always located within the calibration station. These products have no part in the buffering function. This means that the workload norm for the rest of the line has to be fixed at K=32 products. This leads to an underestimation of the throughput. The calibration station in the model is starved when the preceding buffer is empty, whereas in reality, the calibration part can keep on working for while, since there are still products at the operating positions.

It necessary to reduce K even further. The throughput time of the products in the model consists only of waiting time, since the product flow is continuous and transportation times are neglected. Therefore the products arrive earlier at the calibration part compared to the real system, which implies that calibration utilization and as a result the throughput is too high. This error can be corrected by subtracting extra products from the workload norm of 32 products in the model. In the case of an empty system, the real throughput time of the pre–assembly is 4 minutes (consisting of cycle times and transportation times) longer than the modeled waiting time. A throughput estimate is the production rate of the calibration part ν (expressed in products per minute), assuming 100%

utilization. Using Little's formula, the extra throughput time in the real system leads to approximately $4 \cdot \nu$ extra products needed in the real system. Hence the workload of the model can be reduced by $4 \cdot \nu$ products. In Table 7.1.2 the resulting workload norm in the model is indicated as a function of the average calibration time.

	average calibration time (sec.)				
	360	390	420	450	480
K	21	22	23	23	24

Table 7.1.2. Workload norm K as function of the average calibration time

In the next section the throughput of the model will be approximated using the method of Chapter 5. The results will be compared with simulation results obtained from a more detailed model. It will be seen that the rough model here used and the approximation method give good results for the throughput.
Three different failure characteristics of the preassembly will be evaluated.

I: average lifetime of 3420 seconds and repair time of 180 seconds, for all assembly stations
II: average lifetime of 1710 seconds and repair time of 90 seconds
III: average lifetime of 855 seconds and repair time of 45 seconds.

These characteristics all lead to the required station availability percentage of at least 95%. All failure rates are assumed to be state–dependent, in the sense that blocked or starved machines cannot fail.

7.1.6 Model analysis

The model described in the previous section is analyzed using the approximation methods of Chapters 4 and 5. In section 5.2 it was remarked that a workload–controlled line with only one integral restriction (that is $K_1=K_2=..K_{N-1}=:K$, notation of section 5.2) is identical to a base–stock–controlled line with a single integral restriction. The line sketched in Fig. 7.1.2 has finite local buffers and a single integral restriction: the total number of products allowed in the line is K. Hence it is possible to apply the approximation method derived for base–stock–controlled lines to the present situation. Subsection 5.1.3 has shown how a continuous–product flow line controlled by the Base–Stock System and with additional local restrictions can be approximated by a locally controlled line. In the notation of section 5.1.3 the local buffer capacities are $C_i=7$, for i=1,..,8 and $C_9=K$. The integral restrictions K_i are all equal to K, for i=1,..,9.

The throughput approximation of the coin–validator line is made in two steps. In the first step, using the method of section 5.1.3, the equivalent local buffer capacities K_i', are determined. In the second step, the resulting line is approximated for the throughput from left to right and from right to left. In this approximation the method described in section 4.3 denoted by AP3 is used (see Table 4.3.7). The final throughput is obtained as the average of the throughput obtained by left–to–right and by right–to–left approximation. By dividing this throughput by the production rate of the calibration part, the line efficiency is obtained.

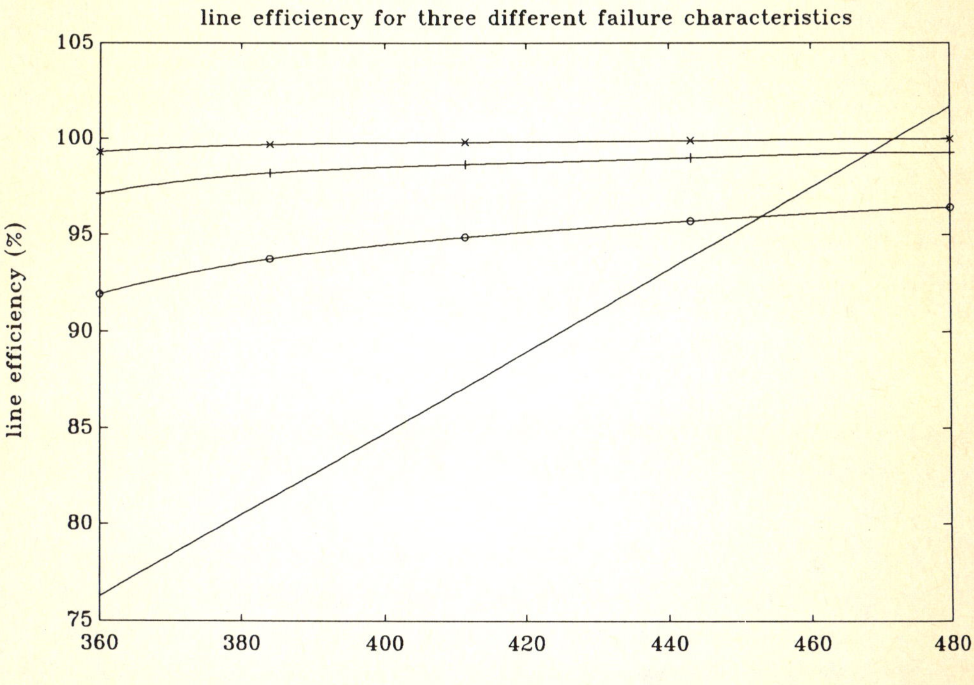

Figure 7.1.3. Obtainable and required line efficiency as a function of the average duration of calibration per product for three different failure characteristics. o: failure characteristic I, +: characteristic II, ×: characteristic III.

In Fig. 7.1.3 the obtained line efficiency (in %) of the coin–validator line as a function of the average calibration cycle time (in seconds) is sketched for three different failure characteristics mentioned in section 7.1.5. The required (annual) line efficiency is also indicated, on the basis of the required average production volume of 122 products per hour. The line efficiency decreases as the average calibration time becomes shorter. This is caused by the fact that the calibration part becomes less of a bottleneck. The overcapacity of the preassembly decreases with decreasing average calibration time. This decrease of overcapacity causes an increase of products in this part of the line, and thus a decrease of products in the calibration buffer. This has a negative effect on the line efficiency. From Fig. 7.1.3 it can be concluded that, if the average calibration duration on an annual base is 420 seconds, then the realized line efficiency is higher than the required line efficiency of 89%. If the annual average calibration duration is longer than 440 seconds, capacity problems may arise. Note, however, that efficiency losses due to waiting for operators at the calibration part are not included. In the case of four operators, such efficiency losses will have hardly any influence. If fewer operators are used, the line efficiency will be lower than indicated in Fig. 7.1.3.

7.1.7 Discussion of results

Some representational errors have been made in the model, the influence of which is discussed below.

1. In the real system the cycle time depends on the number of operating calibration stations. No such distinction is made in the model. This error is *over*–corrected by assuming that, if the local buffer of the calibration part is empty, this station stops producing. In the real production system the calibration part continues producing because all operating positions are still occupied. Therefore the throughput of the model is less than the throughput of the real system.
2. In the model the product is an aggregate: no difference is made between country variants. In the real system there is in principle a possibility that there are products in the calibration part while some calibration stations are idle. This occurs if the waiting products are not of the country type which is currently calibrated at the idle stations. This error is however corrected by the set–up in the real system. Products are set up in sequence of earliest requirements and therefore also generally arrive in that same sequence. Products

which have just arrived at the calibration buffer had the earliest shortage of workload. So these products are the ones which are needed most urgently and hence idleness is prevented. Therefore this representational error has little influence (or none at all) on the throughput.

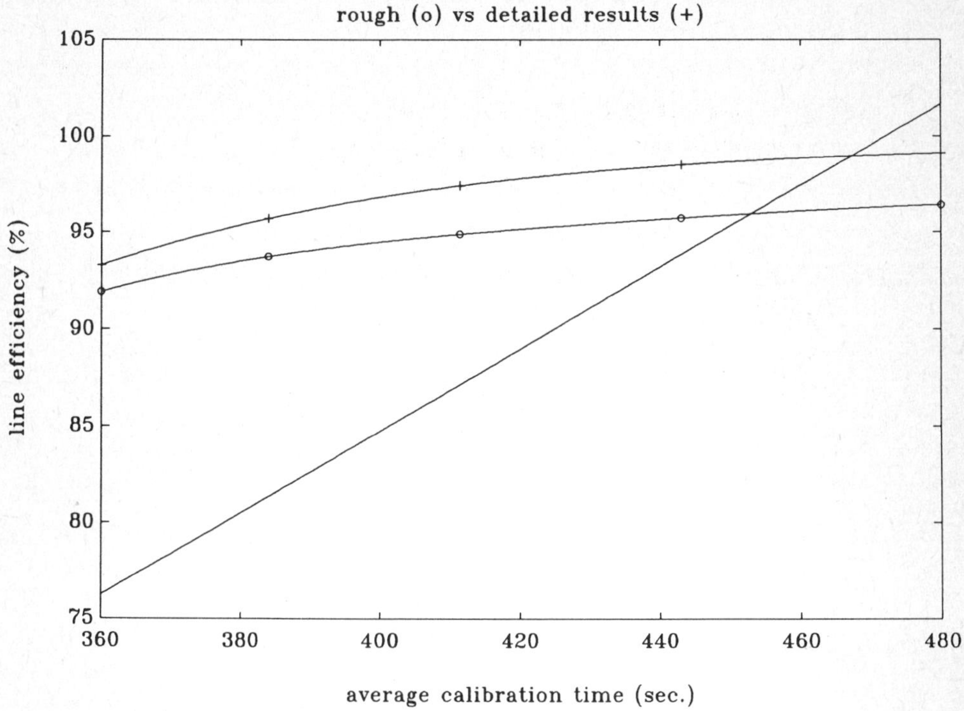

Fig. 7.1.4. Comparison of approximation and detailed simulation (for failure characteristic I) for a single product type

To obtain some insight into the effect of other representational errors, like the assumption of a continuous product flow and aggregation over the products, detailed simulations of the line were carried out. Underlying assumptions in this simulation model are:

1. The product flow is discrete (machines have service times, not production rates).
2. All country variants are explicitly modeled (with different cycle times at specific calibration stations).
3. Failure characteristics and machine speeds are the same as in the model previously discussed. Failures are state–dependent. Cycle times are constant (see Table 7.1.1).

4. The conveyor speed is one product position in two seconds.
5. All machines calibrating a single country variant are clustered. For instance, if there are four country variants, then calibrators 1–4 calibrate variant 1, calibrators 5–8 variant 2, calibrators 9–12 variant 3 and calibrators 13–16 variant 4
6. At assembly station 9, country variants are allocated to the calibrator, dealing with the particular variant, having the smallest remaining workload (expressed in minutes), the product at the operating position incorporated.

This model was simulated for 10,000,000 seconds, starting from an initial run, with only one product type. The results are compared with the approximation results for the earlier model for failure characteristic I, in Fig. 7.1.4. The approximation results indeed appear to be lower estimates for the real line performance. The differences are reasonable.

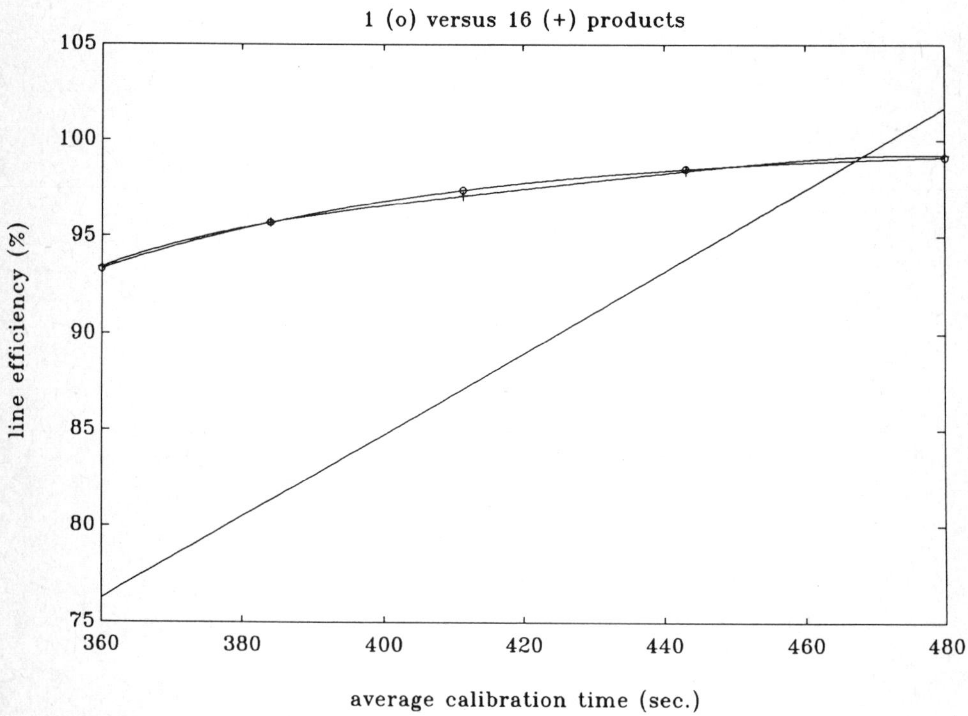

Fig. 7.1.5. Detailed simulations of a single product type and sixteen product types

The fact that only a single product type was simulated in the detailed model has hardly any influence on the (aggregate) throughput. In Fig. 7.1.5 two different simulations of the detailed model are shown. One with a single product type ("ɒ") and one with sixteen different product types ("+"). This sixteen–product–type situation was also simulated for 10,000,000 seconds. Depending on the average calibration time c, the calibrator cycle times c_i are determined as follows. For an average calibration time c of 360 or 480 seconds, all calibrators had cycle times of c_i=360 or c_i=480 seconds, respectively. For c=348.0, c_i=360, i=1,..,8 and c_i=411.4, i=9,..,16. For c=411.4, c_i=360, i=1,..,8 and c_i=480, i=9,..,16. For c=443.1, c_i=411.4, i=1,..,8 and c_i=480, i=9,..,16.

From these figures it can indeed be concluded that even a fairly rough capacity model gives good prediction of the real system performance.

7.2 DIFFERENTIAL–GEAR FACTORY

In this section, the differential–gear factory introduced in section 1.1. will be described, modeled and analyzed at capacity level, using the tools of this text. For several reasons this is a difficult task.

1. Aggregation over products is difficult, since there are many products (4 will be included in the model), many capacities (7 will be included), and the products have different capacity requirement ratios at nearly all capacities.
2. Lot sizes. All product codes have a specific (production) lot size in the pre–operations and a different lot size in the final operations.
3. Manual handling is involved in nearly all operations. Operators can handle several machines simultaneously. In some parts of the factory they work in two shifts, in other parts in three shifts a day.

The emphasis will be more on the type of modeling than on the performance aspect of the case. The modeling is fairly rough. It is shown that reasonable lower bounds for the performance can be obtained. It is possible however, to make the model more accurate, at the cost of increasing the number of machine states. The model will be validated only partially by a detailed simulation of the preoperations and teething capacity groups (which is the bottleneck in the factory). The factory is described in subsection 7.2.1. The performance requirements of the factory are indicated in subsection 7.2.2. The model of the factory is presented in subsection 7.2.3. Capacity control in the model is discussed in subsection 7.2.4. The model is analyzed, using tools of Chapters 4 and 5, in subsection 7.2.5. A comparison with simulation results is given in subsection 7.2.6.

7.2.1 Case description

As mentioned in Chapter 1, the factory produces 65 different cogwheel/bevel–gear–set types. The sets are the major building parts of differential gears. They are supplied to an axle factory, before they are assembled into trucks. In Fig. 1.1.1 a sketch of the main flows in the system is given. For the capacity control we will restrict ourselves to the cogwheel flow, since the bevel gears require separate machines which have a large overcapacity compared with the cogwheel machines. In the cogwheel flow and the part where cogwheels are joined to bevel gears to form sets, only seven capacity groups are distinguished, some of which consist of parallel machines. In fact there are more groups in the cogwheel flow: four in the pre–operations, four in teething, ten in heating, one in grinding and three in polishing/ testing. However, the other capacity groups have such a large overcapacity that they do not influence the throughput. These other capacity groups may influence the throughput time. However, since the system is workload–controlled in the model, by keeping the total workload constant, at a proper norm, it follows from Little's formula, that a good estimate of the throughput automatically yields a good estimate of the throughput time. In this workload norm in the model, the average contents of noncritical capacity groups, not included in the model, have to be incorporated.

The 65 sets of cogwheel/ bevel gears are made up of 30 different cogwheel codes. These codes can be divided into four main types (A, B, C and D). The flow over the different capacity groups is sketched in Fig. 7.2.1.

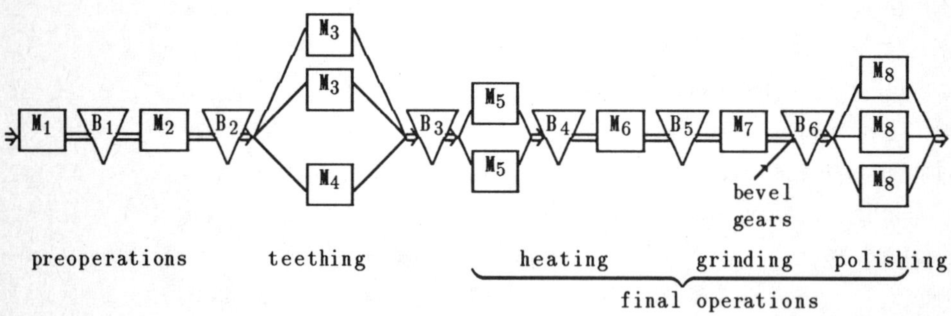

Figure 7.2.1. The considered capacity groups in the differential–gear factory

All cogwheel codes have the same routing over the above–mentioned seven capacity groups. Machine i is denoted by M_i. Machines with the same notation in Fig. 7.2.1 are identical and work parallel. The machines of type M_3 and machine M_4 also work parallel. Parallel machines are counted as being a single capacity group. In Table 7.2.1 the cycle times and switch–over times (separated by a slash) are given for all cogwheel codes at the different capacity groups. The time unit is periods, an hour containing 200 periods.

		pre–operations		teething		heating		grinding	polishing			yearly demand
Type	code	M_1	M_2	(2x)M_3	M_4	(2x)M_5	M_6	M_7	(3x)M_8	Q^1	Q^2	(87/88)
A	1	14.6/500	22.6/200	32/100	112/100	50.18	18.35	7.2/380	22/180	152	8	1960
	2	14.6/500	22.6/200	32/100	112/100	50.18	18.35	7.2/380	22/180	104	8	532
	3	14.6/500	22.6/200	32/100	112/100	43.01	18.35	7.2/380	22/180	360	24	12456
												14948
B	4	11.96/500	14/25	30/100	97.5/100	30.11	14.4	7.2/380	55/180	54	18	144
	5	11.96/500	14/25	30/100	97.5/100	30.11	14.4	7.2/380	55/180	108	18	510
	6	10.96/500	14/25	30/100	97.5/100	30.11	14.4	7.2/380	45/180	126	18	1032
	7	11.96/500	14/25	30/100	97.5/100	30.11	14.4	7.2/380	45/180	216	18	2730
	8	11.81/500	14/25	30/100	97.5/100	30.11	16.28	7.2/380	55/100	108	18	620
	9	8.24/500	14/25	30/100	97.5/100	30.11	14.4	7.2/380	55/100	252	18	2990
	10	11.96/500	14/25	30/100	97.5/100	30.11	14.4	7.2/380	45/100	90	18	460
												8486
C	11	12.94/500	8/50	318/100	76/100	21.51	10.58	7.2/380	46.2/180	108	12	560
	12	11.36/500	8/50	318/100	76/100	21.51	10.58	7.2/380	46.2/180	132	12	1394
	13	12.94/500	8/50	318/100	76/100	21.51	10.58	7.2/380	54.0/180	60	12	210
	14	11.47/500	8/50	318/100	76/100	21.51	10.58	7.2/380	46.2/180	36	12	74
	15	11.36/500	8/50	371/100	69/100	21.51	10.58	7.2/380	71.2/180	60	12	258
												2496
D	16	12.94/500	16.5/50	221/100	45/100	21.51	8.47	7.2/380	52.1/180	108	18	746
	17	12.94/500	16.5/50	221/100	45/100	21.51	8.47	7.2/380	52.1/180	156	18	1888
	18	9.32/500	16.5/25	221/100	45/100	18.82	9.55	10.9/380	52.1/180	144	18	1594
	19	9.32/500	16.5/25	221/100	45/100	18.82	9.55	10.9/380	52.1/180	53	18	220
	20	9.32/500	16.5/25	221/100	45/100	18.82	9.55	10.9/380	52.1/180	36	18	144
	21	9.32/500	16.5/25	221/100	45/100	18.82	9.55	10.9/380	52.1/180	108	18	778
	22	9.32/500	16.5/25	221/100	45/100	18.82	9.55	10.9/380	52.1/180	108	18	726
	23	9.23/500	16.5/25	221/100	45/100	16.13	9.55	10.9/380	42.1/180	54	18	150
	24	9.23/500	16.5/25	221/100	45/100	16.13	9.55	10.9/380	42.1/180	54	18	156
	25	9.14/500	16.5/25	221/100	45/100	16.13	9.55	10.9/380	33.8/180	18	18	50
	26	9.14/500	16.5/25	221/100	45/100	16.13	9.55	10.9/380	33.8/180	36	18	62
	27	9.14/500	16.5/25	221/100	45/100	14.12	9.55	10.9/380	35.7/180	72	18	364
	28	9.14/500	16.5/25	221/100	45/100	14.12	9.55	10.9/380	35.7/180	90	18	418
	29	9.14/500	16.5/25	221/100	45/100	14.12	9.55	10.9/380	35.7/180	108	18	548
	30	9.14/500	16.5/25	221/100	45/100	14.12	9.55	10.9/380	35.7/180	108	18	600
												8444
												34374

Table 7.2.1. Data, including cycle times/switch–over times (in periods) for the different capacity groups. Q^1 is the lot size in preoperations, Q^2 is the lot size in final operations.

The cycle times are in general not constant, but show some variability. These variations in cycle times are mainly due to set–ups of products by operators. The switch–over times are the times needed to switch over from a code to an other arbitrary code (also within the main types A,B,C and D).
Since all codes listed in Table 7.2.1 require time to be set up, production is in batches. For all product codes there are two different production lot sizes. There is a code–dependent lot size in preoperations and teething and another (smaller) lot size in heating, grinding and polishing. The lot sizes in the preoperations are derived from the annual demand and Camp's formula. However, they are adjusted so that they are coordinated with lot sizes in heating and polishing. A lot size in preoperations is *n* times the lot size in final operations. The lot sizes in the final operations are based on transportation lots. The lot size depends on the number of parts that a transportation box can contain. Before heating (M_5), the parts are loaded in such boxes. Usually a complete preoperation lot is dealt with before another one is loaded.

Although production is in batches in preoperations, there are no transportation batches: as soon as M_1 finishes an item, M_2 can start operating on it (provided M_2 is set–up correctly). The buffer B_1 is FCFS (First Come First Served) emptied. Operators are needed to set up products at M_1 and M_2. A single operator controls both machines. In this part of the factory, work is done in three shifts of 7.5 hours each per day.

The next downstream capacity group is teething. There are three parallel teething machines. Two teething machines are identical, of type M_3. Machine M_4 differs from M_3. All teething machines can handle all product types, but the types A and B have priority on the machines M_3 and the types C and D have priority on machine M_4. This is, because product types A and B can be operated faster by machines M_3, types C and D are are faster operated by machine M_4. See Table 7.2.1. If necessary, all machines can work simultaneously on the same batch. If there are items of types A or B and simultaneously items of type C or D present in the buffer, then machines M_3 produce solely A and B, and machine M_4 produces C and D. Two operators work in this capacity group in two shifts.

In final operations (heating, grinding and polishing) production is in batches and transportation is also in batches. The production–batch size in all capacity groups in final operations (apart from ovens M_5) equals the transportation–batch size (the contents of a transportation box). Since the batch sizes in the pre–operations are larger than in final operations, usually several identical batches are produced subsequently in final operations. Machines M_5 and M_6 are ovens. Both ovens M_5 are carousel ovens in which products are carbonated. Products are unloaded one by one from the transportation box into one of the ovens. Then they are loaded again into transportation boxes. The ovens M_5 each work on an individual batch consisting of the contents of one transportation box. Work is in three shifts a day.

M_6 is a transit oven, which can contain several batches (boxes). Products pass this oven on a conveyor belt. Immediately after oven treatment they are ground by M_7. The cycle times for M_6 in Table 7.2.1 are cycle times for a completely filled oven. If the oven is empty, the cycle time of the first product of a batch is longer, since the oven has to be crossed. Machines M_6 and M_7 are set–up simultaneously, since they are operated by the same operator. In Table 7.2.1 the set–up time is indicated under machine M_7.

Sets of cogwheels and bevel gears are formed before polishing (M_8). It is assumed here that there is no waiting by cogwheels for corresponding bevel gears. This assumption is justified since the bevel–gear machines have a large overcapacity compared to cogwheel machines. Since orders for a set are started simultaneously at the cogwheel and bevel–gear machines, the bevel gears will be present at B_6 before the cogwheels have arrived. The sets are polished at machines M_8. In contrast to the situation at machines M_3 and M_4, a batch is handled by only one of these machines. In grinding and polishing, work is in two shifts a day.

7.2.2 Control objectives

The yearly demand in Table 7.2.1 does not vary much within a few years (apart from long–term trends). Also the type–mix distribution is more or less constant. The code–mix distribution varies more. The (aggregate) yearly assembly plan is based on the expected demand. Its volume is fixed and it is distributed equally over the 13 periods within a year. The plan for one such period (four weeks) contains the detailed needs for individual codes. During a period, no changes are made in the plan.

The aim is to realize the monthly assembly plan and to realize short lead times. Realization of the assembly plan means that at aggregate level a throughput of 34374 products per year has to be realized. Since a year has 250 working days, this leads to a desired throughput of 137.50 products per day. Another aim is to reduce the throughput times to one week. At the moment the lead time is three–

four weeks.

At detailed level it has to be guaranteed that the right products are set up, in order to meet the due dates.

In the next subsection the machines in the factory will be modeled at aggregate level. The choice of the workload norm in the model is considered in subsection 7.2.4. In subsection 7.2.5 it is analyzed.

7.2.3. Modeling the case

In this subsection an aggregate model of the system is built. It is meant to illustrate the use of the continuous–flow model in the present situation. The aim is to determine whether the work–in–process can be reduced, while maintaining the same throughput. In modeling the system the following assumptions are made:

1. Operators are assumed to be present when needed. In reality this is in fact a target set by management. Operators that have to fulfill other tasks, besides controlling cogwheel machines, also controlling bevel–gear machines, for instance, have to make sure that the other tasks do not cause idleness of cogwheel machines. At M_1 and M_2, where this assumption is violated, the operator still has to try to let the slowest of the two machines (M_2) not be starved.
2. Raw materials (casts) are assumed to be present when needed. In reality this has to be achieved not by purchasing the raw materials on the basis of the assembly plan, but on the basis of a derived plan, which has a longer horizon. This is necessary because the procurement times may vary up to fifteen weeks, which is far beyond the horizon of the assembly plan.
3. At capacity level the factory is controlled by workload control. The total number of products in the factory is kept at a certain level. The order release is product by product (within a batch) and not batch by batch.
4. It is assumed that the machines do not fail. In reality some machines do fail, but failure characteristics are unknown. However, it is not difficult to include failures in the model, if so desired.

In modeling the aggregate goods flow over the different capacity groups, we proceed in two different steps.

1. The codes within a type are aggregated.
2. The types are aggregated.

As far as step 1 is concerned, it can be seen from Table 7.2.1 that, within a type, there are only slight differences in cycle times and switch-over times at all capacities. The cycle time of the type (an aggregate) at a capacity is determined to be the weighted cycle time of all codes within the type. That is, if r_i is the relative fraction of yearly demand of code i within the type, then the cycle time is determined as $\sum_i c_i r_i$, with c_i the cycle time of code i. The switch-over time T and batch-size Q are determined similarly: $T=\sum_i T_i r_i$ and $Q=\sum_i Q_i r_i$. The errors made in this way in the switch-over times and cycle times are slight, in view of the slight differences between codes. The errors in batch sizes in preoperations are greater. This will (partially) be compensated later by assuming that the processing time of a complete batch is a random variable. The results are given in Table 7.2.2.

Type	M_1	M_2	(2x) M_3	M_4	(2x) M_5	M_6	M_7	(3x) M_8	Q^1	Q^2	yearly demand
A	14.6/500	22.6/200	32/100	112/100	44.21	18.35	7.2/380	22/180	323.62	21.3	14948
B	10.5/500	14/25	30/100	97.5/100	30.11	14.54	7.2/380	50/142	193.78	18	8486
C	11.9/500	8/50	323/100	75.3/100	21.51	10.58	7.2/380	49.4/180	110.27	12	2496
D	10.4/500	16.5/33	221/100	45/100	18.45	9.21	9.7/380	47.8/180	117.49	18	8444

Table 7.2.2. Cycle times/switch-over times (in periods) for types

The aggregation over the types (to simplify notation, from now on denoted by 1, 2, 3 and 4, respectively) is as described in section 6.1. The types can be distinguished from each other at each machine, since each type has a particular cycle time. However, in the buffers there is only one product, the aggregate. Here the products cannot be distinguished any more. The solution is to count buffer contents in units and to assume that in each buffer the type mix is identical to the yearly demanded type mix $\mathbf{p}=(14948,8486,2496,8444)/34374$.

The machines are now so modeled, that the type mix within each buffer is maintained. All machines are modeled as in section 2.4, that is with production rates ν_j, exponentially distributed sojourn times in all states, with rates λ_j and a transition matrix $\mathbf{P}$. If a machine produces N types (N=4 for all machines), then the machine has N+1 states (1,2,..,N,N+1). State j (j=1,..,N) corresponds to production of type j (note however, that the machine is not *really* able to distinguish between the products it takes from the buffer). After production of type j, it switches over to production of another product. This switching over is modeled by the transition matrix $\mathbf{P}^i$ for machine M_i. The structure of the transition matrix is

$$\mathbf{P}^i = \begin{bmatrix} 0 & 0 & 0 & 0 & 1 \\ 0 & 0 & 0 & 0 & 1 \\ 0 & 0 & 0 & 0 & 1 \\ 0 & 0 & 0 & 0 & 1 \\ p^i_{51} & p^i_{52} & p^i_{53} & p^i_{54} & 0 \end{bmatrix}. \tag{7.1}$$

From this structure we see that, after production of a certain type, M_i switches over to state 5, with a production rate 0, after which a transition is made to state j with switch–over probability p^i_{5j} (j=1,..,4). Let $\boldsymbol{\pi}^i$ be the equilibrium probability for the machine states of machine M_i. The switch–over probabilities q^i_{5j} are determined by requiring each machine to produce the equilibrium product mix (see section 6.1). That is

$$\pi^i_j = \begin{cases} a^i \dfrac{p_j}{\nu^i_j} & ,j=1,..,N \\ 1-a^i \sum\limits_j \dfrac{p_j}{\nu^i_j} & ,j=N+1 \end{cases} \tag{7.2}$$

for some constant a^i. The quantity $1-a^i\sum\limits_j \dfrac{p_j}{\nu^i_j}$ is the fraction of time that machine M_i is set up. Together with the set of equations (7.3) for the equilibrium probabilities and the additional condition $\sum_j p^i_{N+1,j} = 1$, the constant a^i and the switch–over probabilities are determined completely.

$$\begin{aligned} \pi^i_j\lambda^i_j &= p^i_{N+1,j}\pi^i_{N+1}\lambda^i_{N+1} \quad ,j=1,..,N \\ \pi^i_{N+1}\lambda^i_{N+1} &= \sum_{j=1}^{N} \pi^i_j\lambda^i_j \end{aligned} \tag{7.3}$$

We will now treat the exact choice of the machine parameters in the model, starting with machines M_1, M_2, M_6 and M_7. These machines all have five states. Then the groups of parallel machines M_5 and M_8 are treated, and finally the combination of machines M_3 and M_4.

1. Modeling M_1, M_2, M_6 and M_7

Machines M_1, M_2, M_6 and M_7 have five states. The first four states correspond to production of the four types. State 5 corresponds to switch–overs (or repairs). The (constant) speeds corresponding to these states are expressed in products per day, using the fact that M_1, M_2 and M_7 work in two shifts and M_6 in three shifts a day. The choice of "days" as time unit enables problems of different shifts to be avoided, at the cost of loss of some detail. It is also possible to use hours as time unit and explicitly include shifts. This can be done by modeling shifts as a special kind of failures. See the discussion in section 7.2.6. The production rates of these machines are obtained from Table 7.2.2. Let ν_j be the production rate of type j (j=1,..,4) and Q_j the mean production lot–size of type j (for M_1, M_2 the pre–operations lot size, for M_6, M_7 the final operations–lot size). The quantity Q_j/ν_j is the mean time needed to produce a batch of type j. The mean sojourn time in states 1,..,4 is determined as Q_j/ν_j. The mean sojourn time for state 5 can be obtained directly from Table 7.2.2 as the mean switch–over time (expressed in days). The distribution for the sojourn times is taken exponential.

Two errors, which are discussed below, are made in these parameter choices. By the first error the system throughput is slightly overestimated. By the second error the throughput is underestimated.

1. Machine failures are not incorporated. Failures can be incorporated by proper choices of the sojourn time distributions in the machine states and the transition matrices. See the case of the insertion department in section 2.1 for an example.
2. In reality, the time to produce a batch shows less variability than the exponential distributions. Exact distributions are not known, not even variances.

The transition matrix $\mathbf{P^i}$, for i=1,2 (corresponding to machines M_1, M_2) has a structure as in (7.1). For M_6 and M_7 the structure of $\mathbf{P^i}$ is similar, except that $p^i_{jj} \neq 0$, i=6,7; j=1,..,4. This is due to the fact that a number of batches of the same type are produced before switching over to another type. The structure of $\mathbf{P^i}$ is then (i=6,7)

$$\mathbf{P^i} = \begin{bmatrix} p^i_{11} & 0 & 0 & 0 & 1-p^i_{11} \\ 0 & p^i_{22} & 0 & 0 & 1-p^i_{22} \\ 0 & 0 & p^i_{33} & 0 & 1-p^i_{33} \\ 0 & 0 & 0 & p^i_{44} & 1-p^i_{44} \\ p^i_{51} & p^i_{52} & p^i_{53} & p^i_{54} & 0 \end{bmatrix} \tag{7.4}$$

The probability $1-p^i_{jj}$ is determined as the quotient Q^2_j/Q^1_j, where Q^1_j is the batch size for type j in preoperations and Q^2_j is the batch size for type j in final operations (see Table 7.2.2). The probabilities q_j^i are determined by substituting equations (7.2) in the system (7.5), and by furthermore requiring $\Sigma_j p_{5j}^i = 1$. Equations (7.5) are the balance equations for the machine states.

$$(1-p_{jj}^i)\pi_j^i\lambda_j^i = p_{5j}^i \pi_5^i \lambda_5^i \qquad ,j=1,..,4$$

$$\pi_5^i\lambda_5^i = \sum_{j=1}^{4} (1-p_{jj}^i)\pi_j^i\lambda_j^i \tag{7.5}$$

2. Modeling M_5 and M_8

The machines M_5 and M_8 are available in multiples. The modeling of parallel machines was treated in section 2.4 (see section 2.4.2). The present situation differs from that section since there are several product types. In principle, the machines M_5 and M_8 can be modeled in a similar way as the above–described machines, with five states per machine. The liferates can be determined by assuming that every machine finishes a complete batch, before it switches–over. The so–modeled machines can then be combined to form a single production unit, with many states. The number of states of such a production unit is quite large (5^2 for M_5 and 5^3 for M_8). In the approximation to follow in subsection 7.2.5 this implies that a system of about 5^3 (in the case of M_5) or about 5^4 (in the case of M_8) linear equations has to be solved (the extra factor 5 for the case of M_5 comes from the fact that the 5^2 states states of M_5 have to be combined with the states of a neighboring aggregate machine, in an approximation sequence (see section 2.6 and Chapter 4). Although this will not lead to insuperable problems, a simpler approximation is chosen here. In this approach, the cycle times as listed in Table 7.2.2, are halved for M_5 and reduced to a third for M_8. The sojourn time in the switch–over state is also halved for M_5 and reduced to a third for M_8. The efficiency of the group of parallel machines as a whole remains the same, but the variance in speed will be greater than in the case of modeling machine M^5 as a machine with 5^2 states (this can be seen rapidly, when considering the somewhat simpler case of two parallel, identical unreliable machines with exponential life and repair times). In the proposed model, all machines are assumed to work simultaneously and they are also simultaneously switched over. Hence the jumps in production rates will be greater than in the case of modeling machines M_5 as a production unit with 5^2 states, and there will also be less jumps per time unit. This will lead to an underestimation of the throughput (in the order of 10%, see also the approximation and simulation results to follow).

The sojourn times are determined in the same way as for machines M_1, M_2, M_6 and M_7. The transition–probability matrix is given by (7.4) for both groups. These groups will be denoted by $\tilde{M}_5$ and $\tilde{M}_8$, respectively.

3. Modeling the combination of M_3 and M_4

Modeling the three parallel machines M_3 (available twice) and M_4, is similar to the modeling of machines M_5 and M_8. They are modeled as a single machine with five states. The corresponding speeds in these states are obtained by summing the speeds of the machines M_3 and M_4. For instance, it can be concluded from figures in Table 7.2.2, that $2{\times}3000/32 + 3000/112$ products of type A can be produced per day (recall that work is in two shifts of 1500 periods a day, where an hour contains 200 periods). The resulting machine is denoted by $\tilde{M}_3$.

The model parameters for all machines in the system are given in Table 7.2.4. Machine speeds are expressed in products per day. Life rates are expressed in $(\text{days})^{-1}$. They are obtained from the figures in Table 7.2.2.

In such modeling, several representational errors are made.

1. Sojourn times in machine states. The sojourn–time distribution in machine states are all chosen as exponential. The average set–up time in the model is chosen equal to the real maximum set–up time for types. By this assumption the throughput is underestimated. For the productive states, the assumption of exponential distributions for the batch–production time also leads to underestimation of the throughput. In preoperations this assumption is justified by the fact that batch sizes show strong variations for codes within a type. In final operations batch sizes are nearly all equal for codes within a type, while cycle times are also nearly constant. It is possible to take nonexponential phase–type distributions for the batch–production time for the final assembly machines, at the cost of increasing the state space of these machines.
2. Halving the cycle times of machines M_3, M_5 and reducing the cycle time of M_8 to a third also leads to underestimation of the throughput, as argued before.
3. An error is made in the aggregation over the products. It is assumed that batches of product types are taken out of each buffer in a random sequence, but so that the yearly product mix is realized. In reality it is possible to choose between the available batches. See also the discussion in section 6.1. It

may be possible to improve the utilization of certain machines by sequencing the batches. Hence the aggregation over the products will in general lead to an underestimation of the throughput.

mach.	M_1	M_2	$\tilde{M}_3$
speeds	(308.22,428.57, 378.15,432.69,0)	(199.12,321.43, 562.5,272.73,0)	(214.29,230.77 58.42,93.82,0)
rates	(0.9524,2.212,3.429, 3.683,9)	(0.6153,1.659,5.101, 2.321,42.91)	(0.6622,1.1909,0.5298, 0.7985,90)
trans. matr.	$\begin{bmatrix} 0 & 0 & 0 & 0 & 1 \\ 0 & 0 & 0 & 0 & 1 \\ 0 & 0 & 0 & 0 & 1 \\ 0 & 0 & 0 & 0 & 1 \\ .25 & .237 & .123 & .39 & 0 \end{bmatrix}$	$\begin{bmatrix} 0 & 0 & 0 & 0 & 1 \\ 0 & 0 & 0 & 0 & 1 \\ 0 & 0 & 0 & 0 & 1 \\ 0 & 0 & 0 & 0 & 1 \\ .25 & .237 & .123 & .39 & 0 \end{bmatrix}$	$\begin{bmatrix} 0 & 0 & 0 & 0 & 1 \\ 0 & 0 & 0 & 0 & 1 \\ 0 & 0 & 0 & 0 & 1 \\ 0 & 0 & 0 & 0 & 1 \\ .25 & .237 & .123 & .39 & 0 \end{bmatrix}$
eff	299.07	243.49	143.00
mach.	$\tilde{M}_5$	M_6	M_7
speeds	(203.57,298.90, 418.41,487.8,0)	(245.23,309.49, 425.33,488.6,0)	(625,625,625, 463.92,0)
rates	(9.544,16.606, 34.868,27.10,500)	(11.497,17.194, 35.44,27.144,500)	(29.30,34.72,52.08 25.773,11.842)
trans. matr.	$\begin{bmatrix} .93 & 0 & 0 & 0 & .07 \\ 0 & .91 & 0 & 0 & .09 \\ 0 & 0 & .89 & 0 & .11 \\ 0 & 0 & 0 & .85 & .15 \\ .25 & .237 & .123 & .39 & 0 \end{bmatrix}$	$\begin{bmatrix} .93 & 0 & 0 & 0 & .07 \\ 0 & .91 & 0 & 0 & .09 \\ 0 & 0 & .89 & 0 & .11 \\ 0 & 0 & 0 & .85 & .15 \\ .25 & .237 & .123 & .39 & 0 \end{bmatrix}$	$\begin{bmatrix} .93 & 0 & 0 & 0 & .07 \\ 0 & .91 & 0 & 0 & .09 \\ 0 & 0 & .89 & 0 & .11 \\ 0 & 0 & 0 & .85 & .15 \\ .25 & .237 & .123 & .39 & 0 \end{bmatrix}$
eff	274.38	307.72	456.32
mach.	$\tilde{M}_8$		
speeds	(409.09,180.0, 182.19,188.28,0)		
rates	(19.179,10.0,15.182, 10.460,52.749)		
trans. matr.	$\begin{bmatrix} .93 & 0 & 0 & 0 & .07 \\ 0 & .91 & 0 & 0 & .09 \\ 0 & 0 & .89 & 0 & .11 \\ 0 & 0 & 0 & .85 & .15 \\ .250 & .237 & .123 & .390 & 0 \end{bmatrix}$		
	235.92		

Table 7.2.4. Resulting machine parameters

7.2.4 Control of the aggregate flow in the model

To realize a good utilization of the teething–capacity group, the cogwheel flow in the model is controlled by workload. The number of products in the model is kept constant. In Chapter 5 it was argued that it does not matter whether the system is locally controlled or workload–controlled, since the same combination of throughput, work–in–process and throughput time can be realized by both control methods. The reason for choosing workload control, is that a good approximation for the throughput automatically yields a good approximation for the throughput time, by Little's formula. Furthermore it yields a relatively small variance in the throughput time (see section 5.4).
Workorders are released in the model, whenever the number of cogwheels (also in sets) in the production system drops below norm. The workload norm in the model is related to the workload norm in the real system (K) as follows. In Chapter 6 it was observed that, in the case that production series differ in size in subsequent stages, there will often be stock in between those stages, which does not have a real buffering function, but which is present because of the different lot sizes. Here there will not be much stock due to lot–sizing effects, since the lot sizes in preoperations and final operations are coordinated. Furthermore, in final operations there is also some time coordination, since several batches of the same type are usually produced subsequently. Runs for these batches in final operations, can be started as soon as the first transportation batch is available. Therefore, there will be some inventory between preoperations and final operations, but the amount will be not much more than the average transportation batch size. Furthermore, between any two capacities in final operations the average inventory due to lot–sizing will be approximately equal to the average size of the transportation batch.
The average transportation batch size can be calculated as

$$Q = \sum_{i=1}^{4} p_i Q_i^2,$$

where Q_i^2 is the transportation batch size in final operations for product i, and p_i is the fractional yearly demand for type i. This leads to Q = 19 products. In the model sketched in Fig. 7.2.1 there are only 4 stock points, where due to lot–sizing an amount Q is present (B_3, B_4, B_5 and B_6). In the real system there are 11 such stock points. This means that in the real system the average inventory due to lot–sizing L equals 11×19 = 209 units of product.

Now assume that in the real system the workload norm is set to K=344 products, say, or 2.5 days of production. The stock due to lot sizing leads to an increase L of the lower buffer bounds (of the buffers in final operations).
It is assumed that all operating positions are occupied. These products at the operating positions have no real buffer function. They have to be subtracted from the workload norm in the model. Since there are about 20 capacities in the system, coontaining products, another 20 products are subtracted from the workload norm. Hence the workload norm in the model has to be set equal to $K':=K-L-20 = 115$ products, or about 1.5 days of production.

Although our main objective is to demonstrate that modeling with the continuous–flow model is possible for such a complicated system, in the next subsections the performance of the model is investigated. Since the modeling has been quite rough, it may be hoped that the model performance provides indeed a lower bound for the system performance.

7.2.5. Analysis of the model

Since the required daily production is 137.5 product units (based on 34374 set types per year, a year containing on average 250 production days), and the machine efficiency of the teething–capacity group is 143.0 products per day, the utilization of this group has to be at least 96.1%. The relation between the work–in–process norm and the throughput and throughput time is established in Table 7.2.5, using approximation methods from Chapters 4 and 5. The results in this table were obtained in the same way as in the previous section for the coin–validator line. The approximation method comprises two steps. First the approximation method of subsection 5.1.2, for lines with a single integral restriction and no local restrictions, is applied. In this way an approximately equivalent line is obtained, with local restrictions only. In the second step this line is approximated, from left to right, using the AP3 method (see Table 4.3.7). Since all machines have more than two states, state–reduction steps were used in both approximations. Such a state–reduction step consists of the approximation of a two–stage line with machines modeled as in section 2.4, by a simpler two–stage line, with four machine states only. This state–reduction step is the same as applied in section 2.7.

In the approximation method, time–dependent transitions are assumed. That means that a machine that is blocked or starved, is switched over at the same rate as for full–speed production. Note that in reality batches are finished before machines are switched over, so that by this assumption the throughput is again underestimated.

The extra amount of products in the system which has no part in the buffering function (L) leads to an increase in throughput time, equal to $\frac{L}{v(K')}$, where v(K') is the throughput of the (modeled) system, with a workload norm of K'=K–L–20= 115 products. This increase in throughput time is incorporated in the throughput time S(K) in Table 7.2.5.

K'(units)	v(K')(units/day)	S(K) (days)
115	137.2	2.51

Table 7.2.5 Approximation of the differential–gear production line. S(K) is the average throughput time

From the results in Table 7.2.5 it can be concluded that, at aggregate level, a throughput time of 2.5 days is realizable, with a corresponding throughput of 137.2 "products" per day. This leads to 34300 products per year, so that the yearly production volume can approximately be realized in fact. Care has to be taken in interpreting these results. It can be expected that, due to the rough modeling, the throughput is underestimated. The throughput will be underestimated even more by the time–dependent approximation.

7.2.6. Evaluation

The modeling of the differential–gear factory is very crude. In view of the rigorous assumptions made (for instance in determining the aggregate and in the modeling of the teething–capacity group) one may wonder to what extent the results in Table 7.2.5 provide lower bounds for performance of the real system. No attempt will be made here to fully investigate the validity of the model, but the approximation results are compared with detailed simulation results of the pre–operations and teething part of the factory. From Table 7.2.4 it can be seen that this part forms the bottleneck.

The simulation model differs in various aspects from the model presented in subsection 7.2.3.

1. Four types of products are distinguished in the simulation model (types 1, 2, 3 and 4). Products are discrete, cycle times are random variables. The distribution of the cycle times is so–called "finite normal", that is the average of a number of random drawings (6) from a uniform distribution whose mean is the mean cycle time and whose lower and upper bounds deviate 50% from the mean. Switch–over times are constant.
2. Batch sizes (preoperation lots) are constant. Batch sizes are chosen as follows: type A: 373; type B: 212; type C: 62; type D: 211. These batch sizes are chosen so that they have the same ratio as in the yearly production plan, while they do not deviate very much from the real average product–type batch sizes.
3. Products are set up cyclically, in the order D–A–C–B–D–A–C–B–.. . The reason for setting up products cyclically is twofold. First, the yearly production volume of each type is produced. Second, machines M_3 and M_4, which form the bottleneck of the system, are loaded alternately. This prevents idleness of the bottleneck as a whole.
4. The machines M_1 and M_2 work in two shifts. This is modeled by letting these machines fail after a constant time of 2920 periods and having them repaired after 1500 periods. Note that an hour contains 200 periods. The products whose processing has already started are finished first. Note also that in the aggregate model of subsection 7.2.3, shifts could have been modeled in this way. However, since operation and "repair" times are constant, this can only be done at the cost of many exponential phases, which leads to many machine states and increases the complexity.
5. The data in Table 7.2.5 are obtained using a workload of 344 products for the system as a whole. Since the simulation model only includes a small part of the factory, it would not be fair to compare the results of Table 7.2.5 with simulation results, obtained by means of workload control, with a workload of K=344 products. Instead, it is assumed that the buffers B_1 and B_2 can contain a finite number of products. The problem is to determine the capacities of these buffers. In the model of the previous section, the products will not be divided equally over all six buffers (see Fig. 7.2.1). The amount upstream of the teething–capacity group is estimated at about one day of production, the amount downstream is estimated at about one and a half days of production. Therefore the buffer capacities in the simulation model are chosen as

$K_1=K_2=69$. products. This keeps the number of products in the simulation model limited as well.

The simulation results were obtained by simulating the system for 4, 8, 12 and 16 weeks (5 days per week), all starting from an initial run of 3 weeks. These results are given in Table 7.2.6.

run time	4 weeks	8 weeks	12 weeks	16 weeks
throughput (prod./day)	174.31	173.85	173.43	172.52
av. cont B_1	65.31	65.67	65.60	65.47
B_2	53.23	53.34	53.47	53.36
tot.cont.	118.54	119.01	119.07	118.83

Table 7.2.6. Simulation results for detailed four-product model

Note that the throughput of the real system will be lower than the 172.5 products per day following from the simulation results for 16 weeks, since the ovens, and the grinding and polishing capacities are not included. Nevertheless, it can be concluded from the simulation results in Table 7.2.6, that the real system will most likely perform better than the approximation model in subsection 7.2.3. Note also that the mean system content upstream of the teething capacity group is nearly one day of work in the simulation model, which means that the choice of the buffer capacities in the simulation model is reasonable.

EPILOGUE

In this text, an approach is given for capacity oriented modeling, design and analysis of production systems. The main features of the modeling aspect are:

1. Aggregation of different products.
2. Continuous modeling of the product flow.
3. Multistate machines with constant machine speeds in each state.
4. Local or integral buffer control policies.

The main analysis tools with which these models can be evaluated are heuristic analytic methods and simulation. Here heuristic tools are advocated. An approximation method is developed for obtaining capacity oriented performance criteria of production systems. The approximation method consists of three parts.

1. A method to approximate two complex machines or two groups of parallel machines, separated by a buffer, by two simple machines.
2. An aggregation method, which replaces two machines and an intermediate buffer by a single machine.
3. A combination of the two above named steps in a sequence to approximate large networks. This sequence depends on the layout of the system and on the way the system is controlled. For flow lines schemes are given in subsection 4.2.3 (local control) and in subsection 5.1.2 (control with BSS). For more complicated networks no schemes are given, but sequences to solve larger networks are selfevident, as is shown by some examples in Chapter 4 (a/d and buffer-sharing networks).

Three cases are considered (in Chapters 2 and 7). All cases have a flow structure. Furthermore they all consist of many stages and many different products have to be manufactured by the same capacity groups. By means of these cases it has been demonstrated that it is possible indeed to describe the cases at aggregate level using the model of this text. All cases were analyzed, using heuristic analytic tools introduced in Chapters 2, 4 and 5. The cases in Chapter 7 were compared with simulation results and it has been shown that this way of modeling and approximating performance gives good results. The approximation method is fast.

How can the modeling and approximation method be used in practice? It can be used as a fast design tool. This is illustrated by the first case in Chapter 7. Although the available data for this case were limited and unsure, still a prediction could be given concerning the behavior of the system and whether the required performance could be realized or not. Closely related, it can also be used as a tool to evaluate different scenarios. For instance, the effect of reallocating buffers, capacity extensions, decreasing set–up times, different control rules etc. on throughput, average work–in–process and throughput times.

Some attention has been paid to the validity of the model by comparing it with detailed simulations. Little attention has been paid to what extent the model is valid in complex production situations, where besides aggregate targets, also detailed performance requirements play a role (specific items have to be delivered in time). This is closely related to the question of to what extent it is allowed to decompose production control in a capacity and an item control level. For the cases there are indeed indications that the decomposition works good. In the coin–validator line, also the detailed performance is good: all individual products are produced in sufficient amounts. This is realized by choosing an item control structure, that is embedded in the aggregate control structure. At aggregate level the control structure is to release products as soon as the *total* number of products drops below 48; at detailed level, product i is released if the number of products of type i drops below $3n_i$, where n_i is the number of calibrators that calibrate country variant i. Note that $\Sigma_i 3n_i = 48$. In the differential gear factory, something similar can be observed. In the aggregate model (aggregate) products belonging to some batch are released if the total number of products falls below 344 items. This gives a certain aggregate performance. In the detailed simulation model of the pre–operations and teething part, products are released in such a sequence, that the teething capacity group as a whole is loaded well, under the aggregate condition that the total number of products in the system is restricted. Comparison of the two models at capacity level gives similar results.

This yields another application area of the model. It can be used at capacity level to set norms for the lower control level of item control. For instance, in the differential–gear factory the average (aggregate) throughput time is 2.5 days. This can be used in planning the release of specific items, given the due dates. Including some safety time, batches have to be released about 5 days before the due dates.

APPENDIX A

In this appendix the two–stage line in subsection 2.3.3 is analyzed for the cases $\lambda_2=0$ (appendix A.1) and $\nu_1=\nu_2$ (appendix A.2). It is shown how the throughput $v(K) = P/T$, the buffer contents distribution function $P[X\leq L]$ and the mean buffer contents can be calculated in these cases.

A.1 The case $\lambda_2=0$

The system (2.2)–(2.5) reduces to

$$\nu_2 f_2'(x) = \alpha_2(x) - \mu_1 f_2(x) + \mu_1 f_4(x) \qquad \text{(A.1)}$$

$$(\nu_2-\nu_1) f_4'(x) = \alpha_4(x) - \lambda_1 f_4(x) + \lambda_1 f_2(x) \qquad \text{(A.2)}$$

with boundary conditions, for $\nu_1>\nu_2$

$$0 = f_2(0) \qquad \text{(A.1a)}$$

$$0 = \alpha_4(K) - \lambda_1 f_4(K) + \lambda_1 f_2(K) \qquad \text{(A.2a)}$$

and for $\nu_1\leq\nu_2$ (2.2a) has to be replaced by

$$0 = \alpha_4(0) - \lambda_1 f_4(0) + \lambda_1 f_2(0) \qquad \text{(A.2b)}$$

We have to solve

$$C_T = \frac{1}{\mu_1}[\alpha_2(0) + \mu_1 f_4(0)] \qquad \text{(A.3)}$$

It will be clear that the solution is trivial if $\nu_1\leq\nu_2$. Therefore we will restrict ourselves to the case $\nu_1>\nu_2$. Let $w(x) := f_4(x) - f_2(x)$ and $\lambda:=\lambda_1$, $\mu:=\mu_1$ then we obtain from (A.1)–(A.2b)

$$w'(x) = \frac{\alpha_4(x)}{\nu_2-\nu_1} - \frac{\alpha_2(x)}{\nu_2} - \left(\frac{\lambda}{\nu_2-\nu_1} + \frac{\mu}{\nu_2}\right) w(x) \qquad \text{(A.4)}$$

with boundary condition

$$w(K) = \frac{\alpha_4(K)}{\lambda} \qquad \text{(A.4a)}$$

Furthermore

$$C_T = \frac{\alpha_2(0)}{\mu} + w(0) \tag{A.5}$$

For the solution of (A.4)–(A.4a) we assume first that the line is unbalanced, that is, $\nu_2 \neq \frac{\nu_1 \lambda}{\lambda+\mu}$ (this is the net average machine speed, or *efficiency*, of machine 1), or $\rho \neq 0$, where $\rho = \frac{\lambda}{\nu_2-\nu_1} + \frac{\mu}{\nu_2}$.

(a) *The unbalanced case, $\rho \neq 0$*

If $\alpha_i(x) = \alpha_i$, for i=2,4, for all $x \in (0,K)$ (that is independent of x) then the solution of (A.4)–(A.4a) is

$$w(x) = \frac{\alpha_4\nu_2 - \alpha_2(\nu_2-\nu_1)}{\lambda\nu_2+\mu(\nu_2-\nu_1)} + \left[\frac{\alpha_4(K)}{\lambda} + \frac{\alpha_2(\nu_2-\nu_1)-\alpha_4\nu_2}{\lambda\nu_2 + \mu(\nu_2-\nu_1)} \right] e^{\rho(K-x)} \tag{A.6}$$

From (A.5) and (A.6) the cost C_T can be determined

$$C_T = \frac{\alpha_2(0)}{\mu} + \frac{\alpha_4\nu_2-\alpha_2(\nu_2-\nu_1)}{\lambda\nu_2 + \mu(\nu_2-\nu_1)} + \left[\frac{\alpha_4(K)}{\lambda} + \frac{\alpha_2(\nu_2-\nu_1)-\alpha_4\nu_2}{\lambda\nu_2 + \mu(\nu_2-\nu_1)} \right] e^{\rho K} \tag{A.7}$$

By substituting the α_i's listed in Table 2.3.2 in (A.7) the corresponding line quantities of Table 2.3.1. can be obtained. For the throughput we obtain

$$\begin{aligned} v(K) &= P/T \\ &= \frac{\nu_2\mu(\nu_1\lambda + (\lambda+\mu)(\nu_2-\nu_1)\, e^{\rho K})}{(\lambda+\mu)(\lambda\nu_2+\mu(\nu_2-\nu_1)e^{\rho K})} \end{aligned} \tag{A.8}$$

For the determination of the buffer content distribution we have to determine PL. By substituting the corresponding α_i's from Table 2.3.2 in (A.4)–(A.5) we obtain the following differential equation

$$\begin{aligned} w'(x) &= \frac{1}{\nu_2-\nu_1} - \frac{1}{\nu_2} - \rho w(x), \quad && x \geq L \\ w'(x) &= -\rho w(x) \quad , && x<L \end{aligned} \tag{A.9}$$

$$w(K) = \frac{1}{\lambda} \tag{A.9a}$$

with costs

$$PL = w(0) \qquad , L>0 \tag{A.10}$$

$$PL = \frac{1}{\mu} + w(0) \qquad , L=0 \tag{A.11}$$

Solution of the system (A.9)–(A.9a), using that w is continuous in L, yields

$$w(x) = \frac{\lambda\nu_1 e^{\rho(L-x)}+(\lambda+\mu)(\nu_2-\nu_1)e^{\rho(K-x)}}{\lambda(\lambda\nu_2+(\nu_2-\nu_1)\mu)} \qquad , x\leq L \tag{A.12}$$

Together with (A.10),(A.11) and (A.7) this implies

$$P[X\leq L] = (T-PL)/T$$

$$= \frac{\lambda[\nu_2(\lambda+\mu)-\mu\nu_1 e^{\rho L}]}{(\lambda+\mu)[\nu_2\lambda+\mu(\nu_2-\nu_1)e^{\rho K}]}, \qquad L<K \tag{A.13}$$

$$P[X=K]= \frac{\mu[\nu_2\lambda+\mu(\nu_2-\nu_1)]e^{\rho K}}{(\lambda+\mu)[\nu_2\lambda+\mu(\nu_2-\nu_1)e^{\rho K}]} \tag{A.14}$$

The solution for the mean buffer content can be obtained by the corresponding substitution for $\alpha_i(x)$, as listed in Table 2.3.2, in (A.4)–(A.5). It can also be obtained by calculating $EX = {}_0\int^K x dP[X\leq x] + KP[X=K]$. The solution is

$$EX = \frac{\mu(\nu_2-\nu_1)\{-\lambda\nu_1\nu_2+[(\lambda+\mu)(\lambda\nu_2+\mu(\nu_2-\nu_1))K+\nu_1\nu_2\lambda]e^{\rho K}\}}{(\lambda+\mu)(\lambda\nu_2+\mu(\nu_2-\nu_1))\,[\lambda\nu_2+\mu(\nu_2-\nu_1)e^{\rho K}]} \tag{A.15}$$

(b) *The balanced case*, $\rho=0$

In case $\rho=0$ equation (A.4) reduces to

$$w'(x) = \frac{\alpha_4(x)}{\nu_2-\nu_1} - \frac{\alpha_2(x)}{\nu_2} \tag{A.16}$$

The line quantities in Table 2.3.1 can be calculated for this case by solving (A.16), with (A.4a) and substituting the solution in (A.5), but also by letting $\rho\to 0$ in the solutions for the case $\rho\neq 0$. We obtain, for the case $\alpha_i(x)=\alpha_i$, $0<x<K$,

$$C_T = \frac{\alpha_2(0)}{\mu} + \frac{\alpha_4(K)\nu_2}{\mu(\nu_1-\nu_2)} + \frac{(\alpha_2(\nu_2-\nu_1)-\alpha_4\nu_2)K}{\nu_2(\nu_2-\nu_1)}, \tag{A.17}$$

$$v(K) = \frac{\nu_2(\nu_2^2+\mu\nu_1 K)}{\nu_1(\nu_2+\mu K)} \tag{A.18}$$

$$P[X\leq L] = \frac{\nu_2(\nu_1-\nu_2)+\nu_1\mu L}{\nu_1(\nu_2+\mu K)}, \quad L<K \tag{A.19}$$

$$P[X=K] = \frac{\nu_2^2}{\nu_1(\nu_2+\mu K)} \tag{A.20}$$

$$EX = \frac{K(\nu_1\mu K+2\nu_2^2)}{2\nu_1(\nu_2+\mu K)} \tag{A.21}$$

The two–stage line with the second machine perfect was also investigated by Meyer et al. (1979). They used a different method to obtain throughput, and buffer content distribution. There appears to be an error in the expressions they obtain for $P[X\leq L]$ and EX (the constant a is wrong). Furthermore in that paper the case $\rho=0$ was not distinguished.

A.2 The case $\nu_1=\nu_2$

In this section we suppose $\lambda_1\neq 0$, $\lambda_2\neq 0$ and $\nu_1=\nu_2$. Equation (2.5) now holds for x=0 and x=K as well, and becomes an ordinary, linear equation. By substituting (2.4) and (2.5) in (2.2)–(2.3) and letting $\nu_1=\nu_2=:\nu$ we obtain

$$-f_1'(x) = \epsilon(x) + r_f[f_2(x)-f_1(x)], \quad 0\leq x<K, \tag{A.22}$$

$$f_2'(x) = \eta(x) + r_g[f_1(x)-f_2(x)], \quad 0<x\leq K, \tag{A.23}$$

where $r_f := \frac{1}{\nu}\left[\frac{\lambda_1\mu_2}{\mu_1+\mu_2} + \frac{\lambda_1\mu_2}{\lambda_1+\lambda_2}\right]$, $r_g := \frac{1}{\nu}\left[\frac{\lambda_2\mu_1}{\mu_1+\mu_2} + \frac{\lambda_2\mu_1}{\lambda_1+\lambda_2}\right]$

and $\epsilon(x) := \frac{\alpha_1(x)}{\nu} + \frac{\lambda_1\alpha_3(x)}{\nu(\mu_1+\mu_2)} + \frac{\mu_2\alpha_4(x)}{\nu(\lambda_1+\lambda_2)}, \quad 0\leq x\leq K$

$\eta(x) := \frac{\alpha_2(x)}{\nu} + \frac{\lambda_2\alpha_3(x)}{\nu(\mu_1+\mu_2)} + \frac{\mu_1\alpha_4(x)}{\nu(\lambda_1+\lambda_2)}, \quad 0\leq x\leq K$

By letting $w(x) := f_2(x) - f_1(x)$, (A.22)–(A.23) can be rewritten as

$$w'(x) = \epsilon(x) + \eta(x) + (r_f-r_g)w(x), \quad 0<x<K \tag{A.24}$$

with a boundary condition (obtained by substituting (2.4) for x=K and (2.5a) into equation (2.2a))

$$0 = \epsilon(K) + r_f w(K) \tag{A.24a}$$

and cost (using A.1a))

$$C_T = \frac{1}{\lambda_2+\mu_1}\,[\eta(0) - r_g w(0)] \tag{A.25}$$

In order to solve the line for the quantities in Table 2.3.1 we have to make the corresponding substitutions of Table 2.3.2 in the system (A.24)–(A.24a) and solve for (A.25). w(0) can be obtained from (A.24)–(A.24a) by assuming continuity in 0. Let $\rho := r_f - r_g$. As in the previous subsection a distinction has to be made between the cases $\rho=0$ and $\rho\neq 0$. Note that $\rho=0$ if and only if $\frac{\nu\mu_1}{\lambda_1+\mu_1} = \frac{\nu\mu_2}{\lambda_2+\mu_2}$, that is, the line is balanced.

(a) *The unbalanced line,* $\rho\neq 0$

If $\alpha_i(x) = \alpha_i$ for $x \in (0,K)$, then $\epsilon(x)=\epsilon$, $\eta(x)=\eta$, for $x \in (0,K)$. The solution of (A.24)–(A.24a) is then

$$w(x) = -\frac{\epsilon+\eta}{\rho} + \left[\frac{-\epsilon(K)}{r_f} + \frac{\epsilon+\eta}{\rho}e^{-\rho(K-x)}\right], \tag{A.26}$$

from which we obtain

$$C_T = \frac{1}{\lambda_2+\mu_1}\,[\eta(0) + \frac{r_g(\epsilon+\eta)}{\rho} - r_g\left[\frac{-\epsilon(K)}{r_f} + \frac{\epsilon+\eta}{\rho}e^{-\rho K}\right] \tag{A.27}$$

From (A.27), Tables 2.3.1 and 2.3.2, P and T can be calculated, which gives

$$v(K) = \frac{\nu\mu_1\mu_2}{(\lambda_1+\mu_1)(\lambda_2+\mu_2)}\left\{1 + \frac{\lambda_2}{\mu_2}\left[1 - \frac{\lambda_1/\mu_1-\lambda_2/\mu_2}{(\lambda_1/\mu_1)e^{\rho K}-\lambda_2/\mu_2}\right]\right\} \tag{A.28}$$

Note that $\nu\mu_1\mu_2/[(\lambda_1+\mu_1)(\lambda_2+\mu_2)]$ is the average line production rate for K=0.
For the calculation of PL (see Table 2.3.1) the system becomes, after substituting $\alpha_i(x)=1$ for all i, $x\geq L$ and $\alpha_i(x)=0$, $x<L$

$$\begin{aligned} w'(x) &= \epsilon^*+\eta^* + (r_f-r_g)w(x), \quad x\geq L \\ w(K) &= -\frac{\epsilon^*}{r_f} \\ w'(x) &= (r_f-r_g)\,w(x), \qquad x<L \end{aligned} \tag{A.29}$$

where $\epsilon^* = \frac{1}{\nu}\left(1 + \frac{\lambda_1}{\mu_1+\mu_2} + \frac{\mu_2}{\lambda_1+\lambda_2}\right)$, $\eta^* = \frac{1}{\nu}\left(1 + \frac{\lambda_2}{\mu_1+\mu_2} + \frac{\mu_1}{\lambda_1+\lambda_2}\right)$

The solution of (A.29) is

$$w(x) = \frac{-(\epsilon^*+\eta^*)}{r_f-r_g}\,e^{-(r_f-r_g)(L-x)} + \left[\frac{-\epsilon^*}{r_f} + \frac{\epsilon^*+\eta^*}{r_f-r_g}\right]e^{-(r_f-r_g)(K-x)}, \quad x<L \tag{A.30}$$

The corresponding cost is

$$PL = \frac{-r_g}{\lambda_2+\mu_1} w(0), \qquad L>0$$

$$= \frac{1}{\lambda_2+\mu_1}\left[\frac{r_g(\epsilon^*+\eta^*)}{r_f-r_g} e^{-(r_f-r_g)L} - r_g(\frac{-\epsilon^*}{r_f} + \frac{\epsilon^*+\eta^*}{r_f-r_g})e^{-(r_f-r_g)K}\right] \tag{A.31}$$

From (A.31) and the mean cycle length T, obtained from (A.27) the buffer content distribution can be obtained

$$P[X\le L] = (T-PL)/T$$

$$= \frac{\eta^* - r_g(\frac{\epsilon^*+\eta^*}{r_f-r_g})(e^{-(r_f-r_g)L}-1)}{\eta^* + r_g(\frac{\epsilon^*+\eta^*}{r_f-r_g}) - \frac{r_g}{r_f}(\frac{\epsilon^* r_g+\eta^* r_f}{r_f-r_g})e^{-(r_f-r_g)K}}, \quad L<K \tag{A.32}$$

$$P[X=K] = \frac{\frac{r_g\epsilon^*}{r_f} e^{-(r_f-r_g)K}}{\frac{r_f\eta^*+r_g\epsilon^*}{r_g-r_f}(\frac{r_g}{r_f} e^{-(r_f-r_g)K}-1)} \tag{A.33}$$

For the calculation of Y we have to substitute $\alpha_i(x) = x$, for all i and $x \in [0,K]$ (see Table 2.3.2). The solution of (2.24)–(A.24a) is then

$$w(x) = -\frac{\epsilon^*+\eta^*}{\rho}(x + \frac{1}{\rho}) + [\frac{-\epsilon^* K}{r_f} + \frac{\epsilon^*+\eta^*}{\rho}(K + \frac{1}{\rho})]e^{-\rho(K-x)}, \tag{A.34}$$

where ϵ^* and η^* are defined as before. This leads to

$$Y = \frac{1}{\lambda_2+\mu_1}[\frac{r_g(\epsilon^*+\eta^*)}{\rho^2} - r_g(\frac{-\epsilon^* K}{r_f} + \frac{\epsilon^*+\eta^*}{\rho}(K + \frac{1}{\rho}))e^{-\rho K}] \tag{A.35}$$

From (A.35) and T, EX can be calculated

$$EX = [\frac{r_g(\epsilon^*+\eta^*)}{\rho^2} - r_g(\frac{-\epsilon^* K}{r_f} + \frac{\epsilon^*+\eta^*}{\rho}(K+\frac{1}{\rho}))e^{-\rho K}] \cdot [\frac{r_f\eta^*+r_g\epsilon^*}{r_g-r_f}(\frac{r_g}{r_f}e^{-\rho K}-1)]^{-1} \tag{A.36}$$

(b) *The balanced case,* $\rho=0$

The solution of (A.24)–(A.24a) is now, in case $\alpha_i(x) = \alpha_i$, $0<x<K$,

$$w(x) = (\epsilon+\eta)x - \frac{\epsilon(K)}{r_f} - (\epsilon+\eta)K \tag{A.37}$$

which yields

$$C_T = \frac{1}{\lambda_2+\mu_1}[\eta(0) + \epsilon(K) + r_g(\epsilon+\eta)K] \tag{A.38}$$

For the throughput we obtain from (A.38)

$$v(K) = \frac{\nu\mu_2(\mu_2+r_g(\lambda_2+\mu_2)K)}{(\lambda_2+\mu_2)^2(1+r_gK)} \tag{A.39}$$

For the calculation of P[X≤L] the system (A.24)–(A.24a) reduces to a system like (A.29), but with ρ=0. Solution yields

$$w(x) = (\epsilon^*+\eta^*)L + \frac{-\epsilon^*}{r_f} - (\epsilon^*+\eta^*)K, \quad x<L \tag{A.40}$$

with ϵ^* and η^* defined as before. This leads to

$$PL = \frac{-r_g}{\lambda_2+\mu_1} w(0), \quad L>0 \tag{A.41}$$

$$PL = \frac{1}{\lambda_2+\mu_1}[\eta^*+\epsilon^* - r_g(\epsilon^*+\eta^*)K], \quad L=0 \tag{A.42}$$

From (A.41) and the cycle length Y, obtained from (A.38), we obtain

$$P[X\leq L] = (T-PL)/T$$

$$= \frac{\eta^*+r_g(\epsilon^*+\eta^*)L}{\eta^*+\epsilon^*+r_g(\epsilon^*+\eta^*)K}, \quad L<K \tag{A.43}$$

$$P[X=K] = \frac{\epsilon^*}{\eta^*+\epsilon^*+r_g(\epsilon^*+\eta^*)K} \tag{A.44}$$

In case $\alpha_i(x) = \alpha_i x$, for all $x \in [0,K]$, the solution of (A.24), (A.24a) is

$$w(x) = \frac{(\epsilon^*+\eta^*)x^2}{2} - \frac{\epsilon^*K}{r_f} - \frac{(\epsilon^*+\eta^*)K^2}{2}, \tag{A.45}$$

where ϵ^* and η^* are defined as before. (A.43) yields

$$Y = \frac{1}{\lambda_2+\mu_1}[\epsilon^*K + r_g\frac{(\epsilon^*+\eta^*)K^2}{2}] \tag{A.46}$$

(A.44) and (A.36) imply

$$EX = K\left[\frac{\lambda_2\mu_1 + \mu_2^2}{(\mu_1+\mu_2)(\lambda_2+\mu_2)} + \frac{r_g}{2}K\right](1 + r_gK)^{-1} \tag{A.47}$$

The formulas for the throughput for both the case $\rho=0$ and $\rho\neq0$ were also obtained by Wijngaard (1979) and by Finch, as reported by Koenigsberg (1959).

A.3 The discrete product model

In this subsection the solution of the system of difference equations for the discrete product model, as described in section 2.3.7, is given, for the case T=1 and $p_i(T):=\lambda_i$, $q_i(T):=\mu_i$. The solution is similar to the continuous flow case with $\nu_1=\nu_2=1$. The system of equations in section 2.3.7 can be rewritten as the system (A.22)–(A.23), except that $\nu=1$, K is an integer and $f_i'(x)$ has to be replaced by $f_i(x+1)-f_i(x)$. Let the parameters r_f, r_g, ϵ, η, ϵ^* and η^* be defined as in appendix A.2 (with $\nu=1$). The function w(x) is defined as $f_2(x)-f_1(x)$. Then w(x) satisfies

$$(1-r_f)w(x+1) = \epsilon+\eta + (r_g-1)w(x) \ , \ 0<x<K \tag{A.48}$$

$$w(K) = \frac{-\epsilon(K)}{r_f} \tag{A.48a}$$

The cost function can be written as (A.25). The solution of the system (A.48), (A.48a) is, for the case $r_f-r_g\neq0$ and $\alpha_i(x)=\alpha_i$, for $x\in\{0,..,K-1\}$,

$$w(x) = \left[\frac{-\epsilon(K)}{r_f} + \frac{\epsilon+\eta}{r_f-r_g}\right]\left[\frac{1-r_g}{1-r_f}\right]^{-(K-x)} + \frac{\epsilon+\eta}{r_g-r_f}, \tag{A.49}$$

In case $\alpha_i(x)=x$, for all i and x, the solution is

$$w(x) = \left[\frac{-\epsilon^*(K)}{r_f} + \frac{\epsilon^*+\eta^*}{r_f-r_g}\left(K + \frac{1-r_g}{r_f-r_g}\right)\right]\left[\frac{1-r_g}{1-r_f}\right]^{-(K-x)} - \frac{\epsilon^*+\eta^*}{r_f-r_g}\left[x + \frac{1-r_g}{r_f-r_g}\right] \tag{A.50}$$

Note that the formulas in (A.49) and (A.50) are close to the formulas for the continuous flow case, namely (A.26) and (A.34), respectively. The difference is, that $e^{-(r_f-r_g)}$ is replaced by $\frac{1-r_f}{1-r_g}$ in (A.49) and (A.50). Furthermore $\frac{1}{r_f-r_g}$ is replaced by $\frac{1-r_g}{r_f-r_g}$ in (A.50). Note that the first two expressions are equal in first order approximation of e^{-r_f}/e^{-r_g}. The last two expressions are both $O((r_f-r_g)^{-1})$.

APPENDIX B

In this appendix the two-stage line is analyzed for the state dependent failure case, described in subsection 2.3.6.

a. $\lambda_2 = 0$

The notations are as in Appendix A.1. We assume $\nu_1 > \nu_2$ (see also Hontelez, 1985).

If $\rho \neq 0$, then, for $\alpha_i(x) =: \alpha_i$, $x \in (0,K)$.

$$C_T = \frac{\alpha_2(0)}{\mu} + \left[\frac{\nu_1\alpha_4(K)}{\nu_2\lambda} + \frac{\alpha_2(\nu_1-\nu_2)+\alpha_4\nu_2}{\nu_2(\nu_1-\nu_2)\rho}\right]e^{\rho K} - \frac{\alpha_2(\nu_1-\nu_2)+\alpha_4\nu_2}{\nu_2(\nu_1-\nu_2)\rho}, \tag{B.1}$$

From (B.1) the throughput can be obtained, as well as the expected cycle length. For PL we obtain

$$PL = w(0) \quad , L>0$$
$$PL = \frac{1}{\mu} + w(0) \quad , L=0,$$

where

$$w(0) = \frac{\nu_1(\nu_2-\nu_1)\mu e^{\rho K} + \nu_1\nu_2\lambda e^{\rho L}}{\nu_2\lambda(\lambda\nu_2 + \mu(\nu_2-\nu_1))} \tag{B.2}$$

For Y we have

$Y = w(0)$, where

$$w(0) = \frac{\nu_1}{(\nu_1-\nu_2)\nu_2\rho^2} + \left[\frac{\nu_1 K}{\nu_2\lambda} - \frac{\nu_1 K\rho - \nu_1}{(\nu_2-\nu_1)\nu_2\rho^2}\right]e^{\rho K} \tag{B.3}$$

For the case $\rho=0$ we obtain

$$C_T = \frac{\alpha_2(0)}{\mu} + \left[\frac{\alpha_2}{\nu_2} + \frac{\alpha_4}{\nu_1-\nu_2}\right]K + \frac{\nu_1\alpha_4(K)}{\mu(\nu_1-\nu_2)} \tag{B.4}$$

$$PL = w(0) \quad , L>0$$
$$PL = \frac{1}{\mu} + w(0) \quad , L=0,$$

with

$$w(0) = \frac{\nu_1}{\mu(\nu_1-\nu_2)} + \frac{\nu_1(K-L)}{\nu_2(\nu_1-\nu_2)} \tag{B.5}$$

$$Y = \frac{\nu_1 K}{\mu(\nu_1 - \nu_2)} + \frac{\nu_1 K^2}{2\nu_2(\nu_1 - \nu_2)} \tag{B.6}$$

b. $\nu_1 = \nu_2 =: \nu,\ \lambda_2 \neq 0$

Substituting (2.16c) in (2.16) and using (2.3a) yields

$$C_T = \frac{\alpha_2(0)}{\mu_1} + \frac{\alpha_4(0)}{\lambda_1 + \lambda_2} + \frac{\lambda_2}{\lambda_1 + \lambda_2} f_1(0) \tag{B.7}$$

Defining $w(x) := f_2(x) - f_1(x)$ yields, for $\rho \neq 0$ (see appendix A.2)

$$w(x) = \left[\frac{-\tilde{\epsilon}(K)}{\tilde{r}_f} + \frac{\epsilon + \eta}{r_f - r_g}\right] e^{-\rho(K-x)} - \frac{\epsilon + \eta}{r_f - r_g}, \tag{B.8}$$

where $\alpha_i(x) = \alpha_i$. The parameters $\rho, \epsilon, \eta, r_f$ and r_g are defined as in appendix A.2 and

$$\tilde{\epsilon}(K) := \alpha_1(K) + \frac{\mu_2 \alpha_4(K)}{\lambda_1 + \lambda_2}, \quad \tilde{r}_f := \frac{\mu_2 \lambda_1}{\lambda_1 + \lambda_2}$$

By (B.7), (B.8) and (2.3a) it follows

$$C_T = \frac{\alpha_2(0)}{\mu_1} + \frac{\alpha_4(0)}{\lambda_1 + \lambda_2} + \frac{\lambda_2}{\lambda_1 + \lambda_2} \left(\frac{\epsilon + \eta}{r_f - r_g} + \left[\frac{\tilde{\epsilon}(K)}{\tilde{r}_f} - \frac{\epsilon + \eta}{r_f - r_g}\right] e^{-\rho K}\right) \tag{B.9}$$

For PL we have

$$PL = \frac{1}{\mu_1} + \frac{1}{\lambda_1 + \lambda_2} - \frac{\lambda_2}{\lambda_1 + \lambda_2} w(0) \qquad , \text{ if } L = 0$$

$$PL = -\frac{\lambda_2}{\lambda_1 + \lambda_2} w(0) \qquad , \text{ if } L > 0,$$

where

$$w(0) = -\frac{\epsilon^* + \eta^*}{r_f - r_g} e^{-\rho L} + \left[-\frac{\hat{\epsilon}}{\tilde{r}_f} + \frac{\epsilon^* + \eta^*}{r_f - r_g}\right] e^{-\rho K}, \tag{B.10}$$

with ϵ^*, η^* defined as in appendix A.2 and

$$\hat{\epsilon} := \frac{\lambda_1 + \lambda_2 + \mu_2}{\lambda_1 + \lambda_2}$$

Furthermore

$$Y = \frac{\lambda_2}{\lambda_1 + \lambda_2} \left(\frac{\epsilon^* + \eta^*}{\rho^2} + \left[\frac{\epsilon^{\ddagger} \cdot K}{\tilde{r}_f} - \frac{\epsilon^* + \eta^*}{\rho}\left(K + \frac{1}{\rho}\right)\right] e^{-\rho K}\right) \tag{B.11}$$

The corresponding quantities for $\rho = 0$ can be obtained from these formulas by letting $\rho \to 0$.

APPENDIX C

In this appendix the moments of buffer increase, buffer decrease and the expected lengths of an increase and decrease period will be derived for an $M/C_2/1$ queueing system with unbounded buffer. The arrival rate is denoted by λ. Let state i correspond to the i th phase of a two–phase Coxian distribution with rates μ_1 and μ_2 and branching probability π (see Fig. 2.7.4, i=1,2).

f_i = number of arrivals during the service, starting in state i (i=1,2)
g_i = number of departures during the interarrival time, starting in state i.

Then $E(U^i)^n = Ef_1^n$, $E(U^d)^n = r_1Eg_1^n + r_2Eg_2^n$, where r_1 and r_2 are the equilibrium probabilities for state 1 and 2, respectively. The definitions of stock increase and decrease U^i and U^d are given in subsection 2.6.1. We have the following equations (parameters as in subsection 2.8.2).

$$Ef_1 = \frac{\lambda}{\lambda+\mu_1}(1 + Ef_1) + \frac{\pi\mu_1}{\lambda+\mu_1} Ef_2 \tag{C.1}$$

$$Ef_2 = \frac{\lambda}{\lambda+\mu_2}(1 + Ef_2) \tag{C.2}$$

After Ef_1, Ef_2 have been obtained from (C.1), (C.2), they can be used to calculate Ef_i^2:

$$Ef_1^2 = \frac{\lambda}{\lambda+\mu_1}(1 + 2Ef_1 + Ef_1^2) + \frac{\pi\mu_1}{\lambda+\mu_1} Ef_2^2 \tag{C.3}$$

$$Ef_2^2 = \frac{\lambda}{\lambda+\mu_2}(1+ 2Ef_2 + Ef_2^2) \tag{C.4}$$

Ef_1^2, Ef_2^2 can again be used for calculating Ef_i^3:

$$Ef_1^3 = \frac{\lambda}{\lambda+\mu_1}(1+ 3Ef_1 + 3Ef_1^2 + Ef_1^3) + \frac{\pi\mu_1}{\lambda+\mu_1} Ef_2^3 \tag{C.5}$$

$$Ef_2^3 = \frac{\lambda}{\lambda+\mu_2}(1 + 3Ef_2 + 3Ef_2^2 + Ef_2^3) \tag{C.6}$$

A similar system of equations holds for the g_i's:

$$Eg_1 = \frac{\mu_1}{\mu_1+\lambda}\,[(1-\pi)(1 + Eg_1) + \pi Eg_2] \tag{C.7}$$

$$Eg_2 = \frac{\mu_2}{\mu_2+\lambda}\,(1+ Eg_1) \tag{C.8}$$

$$Eg_1^2 = \frac{\mu_1}{\mu_1+\lambda}\,[(1-\pi)(1 + 2Eg_1 + Eg_1^2) + \pi Eg_2^2] \tag{C.9}$$

$$Eg_2^2 = \frac{\mu_2}{\mu_2+\lambda}\,(1 + 2Eg_1 + Eg_1^2) \tag{C.10}$$

$$Eg_2^3 = \frac{\mu_1}{\mu_1+\lambda}\,[(1-\pi)(1 + 3Eg_1 + 3Eg_1^2 + Eg_1^3) + \pi Eg_2^3] \tag{C.11}$$

$$Eg_2^3 + \frac{\mu_2}{\mu_2+\lambda}\,(1 + 3Eg_1 + 3Eg_1^2 + Eg_1^3) \tag{C.12}$$

Solution of equations (C.1)–(C.6) gives

$$E(U^i) = a/(\mu_1\mu_2) \tag{C.13}$$

$$E(U^i)^2 = b/(\mu_1^2\mu_2) \tag{C.14}$$

$$E(U^i)^3 = \frac{\lambda[\mu_1^2\mu_2^2 + 3\mu_2 b + 3\mu_1\mu_2 a + \pi\mu_1^2(\mu_1\mu_2 + 3\lambda(2\lambda+2\mu) + 3\lambda\mu_1)]}{\mu_1^3\mu_2^2} \tag{C.15}$$

where $a = \lambda(\pi\mu_1 + \mu_2)$ and $b = \lambda[\mu_1\mu_2 + 2a + \pi\mu_1(2\lambda+\mu_2)]$

Solution of (C.7)–(C.12) and using $r_1 = \frac{\mu_2}{\pi\mu_1+\mu_2}$, $r_2 = \frac{\pi\mu_1}{\pi\mu_1+\mu_2}$ gives

$$E(U^d) = \mu_1\mu_2/a \tag{C.16}$$

$$E(U^d)^2 = \mu_1\mu_2[c + 2d]/(ac) \tag{C.17}$$

$$E(U^d)^3 = \mu_1\mu_2[c^2 + 3cd + 3d(c+2d)]/(ac^2) \tag{C.18}$$

where $c = \lambda(\lambda+\pi\mu_1+\mu_2)$, $d = \mu_1(\mu_2+(1-\pi)\lambda)$. EL^i equals the average service time

$$EL^i = (\pi\mu_1+\mu_2)/(\mu_1\mu_2) \tag{C.19}$$

EL^d equals the average interarrival time

$$EL^d = \frac{1}{\lambda} \tag{C.20}$$

It may be clear that this method to obtain the buffer moments works for each PH/PH/1 queueing system.

For GI/G/1 queueing systems we have to proceed differently. The buffer increases, for instance, can then numerically be determined as follows. Let

α_i = P[i arrivals during a service]
N(t) = number of arrivals in [0,t]
$P_j(t) = P[N(t) = j]$

Then we have the following recursive relation for the P_j's

$$P_0(x) = 1 - A(x), \tag{C.21}$$

$$P_j(x) = \int_0^x P_{j-1}(x-\xi)dA(\xi), \; j \geq 1 \tag{C.22}$$

where A(·) is the interarrival time distribution function. From the P_j's the α_i's can be determined.

$$\alpha_i = \int_0^\infty P_i(x)dB(x), \tag{C.23}$$

where B(·) is the service time distribution. $E(U^i)^n$ is now determined as

$$E(U^i)^n = \sum_{i=0}^{\infty} i^n \alpha_i \tag{C.24}$$

APPENDIX D

In this appendix it will be illustrated how the machine speeds can be determined in continuous flow buffer–sharing networks as introduced in subsection 3.4.2. The used priority rule is rule R_1, defined in that section. This priority rule determines the machine speeds in buffer–sharing networks uniquely, if no loops or bypasses occur in the network. If a network does contain loops or bypasses, the system of equations for the machine speeds may be inconsistent or have multiple solutions. Suppose that machine M_3 in the network in Fig. 3.4.5, which contains a bypass, has priority over machine M_1. Machine M_2 has priority over machine M_3. In this case the system of equations for the machine speeds, using R_1 is:

$$w(M_1) = \min\{4,2-w(M_3)\}$$
$$w(M_2) = \min\{1,2,w(M_1)\}$$
$$w(M_3) = \min\{3,2,2-w(M_2)\}.$$

This leads to $w(M_2) = \min\{1,2-\min\{2,2-w(M_2)\}\} = \min\{1,\max\{0,w(M_2)\}\}$. Hence, for each $w(M_2) \in [0,1]$, there is a solution of this system of equations.

In order to avoid such situations, it is assumed that the buffer–sharing network and priorities satisfy the following additional conditions.

(1) There is a directed path in the network from a source to a sink buffer.

(2) The network contains no loops (directed cycles).

(3) For every *buffer bypass* (a set of two different directed paths from a buffer $b_1 \in B$ to another buffer $b_2 \in B$) that the network contains, it holds that if the priority of the first arc of the one path is greater than the priority of the first arc of the other path at b_1, then the same has to hold at b_2 for the last arc of both paths.

In order to determine the machine speeds, the following additional variables are needed. With each buffer b_1 that has downstream machines there is associated an output rate restriction $o(b_1) \in [0,\infty]$, which is the maximum possible output rate of products from this buffer. This output rate restriction makes that preceding machines may be blocked. If b_1 is non–empty then $o(b_1) = \infty$. With each buffer b_2

that has upstream machines there is associated an input rate restriction $i(b_2) \in [0,\infty]$, which is the maximum possible input rate into this buffer. The input rate restriction causes that following machines may be starved. If b_2 is non–full then $i(b_2) = \infty$.

It is assumed that the network is connected. If not, each connected component can be treated individually.

```
Algorithm
Initialization for all full buffers b: i(b):=0;
               for all empty buffers b: o(b):=0;
while B ≠ ∅
  do found:=false;
     choose source buffer a;
     while not found
     do while a is not a sinkbuffer and not found
          do go downstream along the arc with the highest priority to
             machine j and further downstream to buffer b;
          if b is not full
          then found:=true;
               w(j):=min{v(j),o(a)}
          else a:=b fi
          od;
          if a has no other incoming arcs with a higher priority than
             the arc along which a was entered and not found
          then found:=true;
               w(j):=min{v(j),i(a),o(predj)} fi;
          while a is not a source buffer and not found
          do go upstream along the arc with the highest priority to
             machine j and further to buffer b;
          if b is not empty
          then found:=true;
               w(j):=min{v(j),i(a)}
          else a:=b fi
          od;
          if a has no other outgoing arcs with a higher priority than
             the arc along which a was entered and not found
          then found:=true;
               w(j):=min{v(j),o(a),i(succj)} fi;
        od;
        remove j from the network with all its arcs and nonconnected
        buffers;
        if j obtained from an empty buffer b then o(b):=o(b)-w(j) fi;
        if j supplied an empty buffer b then o(b):=o(b)+w(j) fi;
        if j obtained from a full buffer b then i(b):=i(b)+w(j) fi;
        if j supplied a full buffer b then i(b):=i(b)-w(j) fi
  od.
```

Figure D.1. Algorithm for machine–speed determination in buffer–sharing networks

In Fig. D.1, an algorithm is described that determines the machine speeds for each combination of buffer contents (either empty, full or non–empty and non–full). At each step, the algorithm locates a machine of which the speed can be calculated and removes this machine with every connecting arc from the network. Non–connected buffers are also removed. The input and output rate restriction of the remaining buffers are adapted, depending on the speed of this particular machine. This procedure is repeated till all machine speeds have been determined. At each removal one or more networks remain which are of the same type as the original network. Let *B* denote the set of buffers, v(j) the potential machine speed of machine j and w(j) the real speed of machine j as influenced by the network. $pred_j$ is the buffer preceding j and $succ_j$ is the buffer succeeding j.

Conditions (2) and (3) guarantee that the algorithm always takes a new machine and a new buffer while going downstream or upstream through the network. Condition (1) guarantees that a sink buffer can be reached, starting from a source buffer. The finiteness of the network guarantees that always a machine can be found of which the speed can be determined. If the remaining network is disconnected then the components are treated one by one. For the production system sketched in Fig. D.2 the algorithm gives the following results, in case buffers B_1 and B_3 are full and buffer B_2 is empty. Denote the source buffer (preceding machine M_1) by buffer B_0, the sink buffer (succeeding machine M_6) by buffer B_4. At each buffer shared by more than one machine, machine M_i has priority over machine M_j if and only if $i<j$.

Initialization: $o(B_0)=i(B_4)=o(B_1)=i(B_2)=o(B_3)=\infty$
$i(B_1)=o(B_2)=i(B_3)=0$.
Machine speed determination: $w(M_6)=1$, $i(B_3)=1$;
$w(M_2)=0.6$, $i(B_1)=0.6$, $i(3)=0.4$;
$w(M_3)=0.3$, $o(B_2)=0.3$, $i(B_1)=0.9$;
$w(M_4)=0.3$, $o(B_2)=0.6$, $i(B_1)=1.2$;
now we have two components left, namely B_0–M_1–B_1 and B_2–M_5–B_3.
$w(M_1)=0.9$, $i(B_1)=0.3$; $w(M_5)=0.4$.

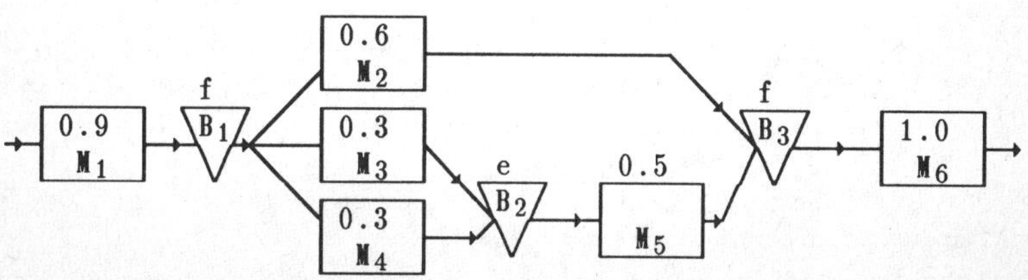

Figure D.2. Buffer–sharing network. The machine speeds are indicated. All buffers are either full (f) or empty (e).

REFERENCES

Akyildiz, I. "Analysis of queueing networks with rejection blocking", *paper presented at First International Workshop on Queueing Networks with Blocking*, Raleigh, 1988a.

Akyildiz, I. "General closed queueing networks with blocking", *Performance '87*, P.–J. Courtois and G. Latouche (eds.), 283–303, Elsevier, Amsterdam, 1988b.

Altiok, T., "Approximate analysis of exponential tandem queues with blocking", *European J. Operational Research 11*, 390–398, 1982.

Altiok, T., "On the phase–type approximations of general distributions", *IIE Transactions 17*, 110–116, 1985.

Altiok, T., "(R,r) production/inventory systems", *Working paper Industrial Engineering*, Rutgers University, 1986a.

Altiok, T., "Approximate analysis for open tandem queues with blocking: exponential and general service times", *IE Working paper 86–111*, Rutgers University, 1986b.

Altiok, T. and H.G. Perros, "Open networks of queues with blocking: Split and merge configurations", *IIE Transactions 18*, 251–261, 1986.

Altiok, T. and H.G. Perros, "Approximate analysis of arbitrary configurations of open queueing networks with blocking", *Annals of Operations Research 9*, 481–509, 1987.

Altiok, T. and R. Ranjan, "Analysis of production lines with general service times and finite buffers: a two–node decomposition approach", *Paper presented at the Fifth International Working Seminar on Production Economics*, Igls, 1988.

Altiok, T. and S. Stidham, "A note on transfer lines with unreliable machines, random processing times, and finite buffers", *IIE Transactions 14*, 125–127, 1982.

Altiok, T. and S. Stidham, "The allocation of interstage buffer capacities in production lines", *IIE Transactions 15*, 292–299, 1983.

Ammar, M.H., "Modeling and analysis of unreliable manufacturing assembly networks with finite storages", *MIT Report* LIDS–TH–1004, 1980.

Ammar, M.H. and S.B. Gershwin, "Equivalence relations in queueing models of manufacturing", *Proceedings of the nineteenth IEEE Conference on Decision and Control*, 715–721, 1981

Ancelin, B. and A. Semery, "Calcul de la productivite d'une ligne integree de fabrication. CALIF, un logiciel industriel base sur une nouvelle heuristique", *Automatique Productique Informatique Industrielle 21*, 209–238, 1987.

Anderson, D.R. and J.C. Moodie, "Optimal buffer storage capacity in production line systems", *International Journal of Production Research 7*, 233–240, 1969.

D'Angelo, H., M. Caramanis, S. Finger, A. Mavretic, Y.A. Phillis and E. Ramsden, "Event–driven model of unreliable production lines with storage", *International J. Production Research 26*, 1173–1182, 1988.

Anick, D., D. Mitra and M.M. Sondhi, "Stochastic theory of a data–handling system", *Bell Systems Technical Journal 61*, 1871–1894, 1982.

Artamonov, G.T., "Productivity of a two–instrument discrete processing line in the presence of failures", *Kibernetika 3*, 126–130 (Engl. tr. *Cybernetics 12*, 464–468, 1977).

Avi–Itzhak, B. and M. Yadin, "A sequence of two servers with no intermediate queue", *Management Science 11*, 553–564, 1965.

Awate, P.G. and B.L.N. Sastry, "Analysis and decompostion of transfer and flow lines – A review", to appear in *Operations Research*, 1987.

Barlow, R.E. and F. Proschan, "*Statistical Theory of Reliability and Life Testing*, Holt, Rinehart and Winston, New York, 1975.

Barten, K., "A queueing simulator for determining optimum inventory levels in a sequential process", *The Journal of Industrial Engineering 13*, 245–252, 1962.

Bechte, W. *"Steuerung der Durchlaufzeit durch belastungsorientierte Auftragsfreigabe bei Werkstattfertigung"*, Dissertation Universitat Hannover, 1980, Fortschrittsgerichte VDI–Z, Reihe 3, Nr. 70, Dusseldorf, 1984.

Beek, P. van, "An application of decomposition and dynamic programming to the N–product, 1–machine problem", *Methods of Operations Research* 41, 63–67, 1981.

Bellman R., *"Introduction to Matrix Analysis"*, McGraw – Hill (2nd ed.), New York, 1970.

Bemelmans, R., *"The Capacity Aspect of Inventories"*, Springer–Verlag, Berlin, 1986.

Bertrand, J.W.M. and J. Wijngaard, "The structuring of production control systems", *International Journal of Operations and Production Management* 6, 5–20, 1986.

Bertrand, J.W.M. and J.C. Wortmann, *"Production Control and Information Systems for Component Manufacturing Shops"*, Elsevier Scientific Pu. Co., Amsterdam, 1981.

Bitran, G.R. and A.C. Hax, "On the design of hierarchical production planning systems", *Decision Sciences* 8, 28–55, 1977.

Bitran, G.R., E.A. Haas and A.C. Hax, "Hierarchical production planning: A single stage system". *Operations Research 29*, 717–743, 1981.

Bitran, G.R., E.A. Haas and A.C. Hax, "Hierarchical production planning: A two–stage system", *Operations Research 30*, 232–251, 1982.

Bocharov, P.P. and F.K. Albores, "On two–stage exponential queueing systems with internal losses or blocking", *Problems of Control and Information Theory 9*, 365–379, 1980.

Bocharov, P.P. and G.P. Rokhas, "On an exponential queueing system in series with blocking", *Problems of control and information theory 9*, 441–455, 1980.

Bolch, G. and I.F. Akyildiz, *"Analyse von Rechensystemen"*, Teubner, Stuttgart, 1982.

Boxma, O. and A. Konheim, "Approximate analysis of exponential queueing systems with blocking", *Acta Informatica 15*, 19–66, 1981.

Brandwajn, A.S. and Y. Lily Jow, "Tandem exponential queues with finite buffers", *Computer Networking and Performance Evaluation*, 245–257, 1986.

Buzacott, J.A., "Automatic transfer lines with buffer stocks", *International Journal of Production Research 5*, 183–200, 1967.

Buzacott, J.A., "Prediction of the efficiency of production systems without internal storage", *International Journal of Production Research 6*, 173–188, 1968.

Buzacott, J.A., "The effect of station breakdowns and random processing times on the capacity of flow–lines with in–process storage", *AIIE Transactions 4*, 308–312, 1972.

Buzacott, J.A., ""Optimal" operating rules for automated manufacturing systems", *IEEE Transactions on Automatic Control 27*, 80–86, 1982.

Buzacott, J.A., "Kanban and MRP controlled production systems", *paper presented at the Fifth International Working Seminar on Production Economics*, Igls, 1988.

Buzacott, J.A. and L.E. Hanifin, "Models of automatic transfer lines with inventory banks; A review and comparison", *AIIE Transactions 10*, 197–207, 1978.

Buzacott, J.A. and D. Kostelski, "Matrix–geometric and recursive algorithm solution of a two–stage unreliable flow line", *IIE Transactions 19*, 429–438, 1987.

Caseau, P. and G. Pujolle, "Throughput capacity of a sequence of queues with blocking due to finite waiting room", *IEEE Transactions Software. Engineering, SE–5*, 631–642, 1979.

Choong, Y.F. and S.B. Gershwin, "A decomposition method for the approximate evaluation of capacitated transfer lines with unreliable machines and random processing times", *IIE Transactions 19*, 150–159, 1987.

Clark, A. and H. Scarf, "Optimal policies for a multi–echelon problem", *Management Science* 6, 475–490, 1960.

Coddington, E.A. and N.Levinson, *"Theory of Ordinary Differential Equations"*, McGraw–Hill, New York, 1955.

Coillard, P. and J.M. Proth, "Effet des stocks tampons dans une fabrication en Ligne" *Revue Belge de Statistique, D'Informatique et de Recherche Operationelle 24*, 3–27, 1984.

Corten, M. and M.B.M. de Koster, "On the reversibility of manufacturing networks", to appear in *J. Mathematical Analysis and Applications*, 1987.

Dallery, Y., David, R. and X–L. Xie, "An efficient algorithm for analysis of transfer lines with unreliable machines and finite buffers", *Tech. Rep. Laboratoire d'Automatique de Grenoble*, 87–17, 1987.

Dallery, Y., David, R. and X–L. Xie, "Approximate analysis of transfer lines with unreliable machines and finite buffers, *report LAG, no. 87–64*, 1988.

Dallery, Y. and Y. Frein, "A decomposition method for the approximate analysis of closed queueing networks with blocking", *presented at First International Workshop on Queueing Networks with Blocking*,Raleigh, 1988.

Dubois, D. and J.P. Forestier, "Productivite et en–cours moyen d'un ensemble de deux machines separees par une zone de stockage", *RAIRO Automatique/Systems Analysis and Control 16*, 105–132, 1982.

Flatto, L., "Two parallel queues created by arrivals with two demands II", *SIAM J. Applied Mathematics 45*, 861–878, 1985

Flatto, L. and S. Hahn, "Two parallel queues created by arrivals with two demands I", *SIAM J. Applied Mathematics 44*, 1041–1053, 1984

Forestier, J.P., "Modelisation stochastique et comportement asymptotique d'un systeme automatise de production", *RAIRO Automatique 14*, 127–144, 1980.

Forrester, J.W., *"Industrial Dynamics"*, Wiley and Sons, New York, 1961.

Foster, F.G. and H.G. Perros, "On the blocking process in queue networks", *European J. Operational Research 5*, 276–283, 1980.

Fox, R.J. and D.R. Zerbe, "Some practical system availability calculations", *AIIE Transactions 6*, 228–234, 1974.

Freeman, M.C., "The effects of breakdowns and interstage storage on production line capacity", *J. of Industrial Engineering 15*, 194–200, 1964.

Gershwin, S.B., "Modeling and analysis of assembly/disassembly networks", MIT *Laboratory for Information and Decision Systems Report*, 1986a.

Gershwin, S.B., "Assembly/Disassembly systems: An efficient decomposition algorithm for tree–structured networks", *MIT report LIDS–P–1579*, 1986b.

Gershwin, S.B., "An efficient decomposition method for the approximate evaluation of tandem queues with finite storage space and blocking", *Oper. Res. 35*, 291–305, 1987.

Gershwin, S.B., "An efficient decomposition algorithm for unreliable tandem queueing systems with finite buffers", *presented at First International Workshop on Queueing Networks with Blocking*,Raleigh, 1988.

Gershwin, S.B. and O. Berman, "Analysis of transfer lines consisting of two unreliable machines with random processing times and finite storage buffers", *AIIE Transactions 13*, 2–11, 1981.

Gershwin, S.B. and I.C. Schick, "Modeling and analysis of unreliable transfer lines with finite insterstage buffers", *MIT–LIDS report* ESL–FR–834–6, 1979.

Gershwin, S.B. and I.C. Schick, "Modelling and analysis of three–stage transfer lines with unreliable machines and finite buffers", *Operations Research 31*, 354–380, 1983.

Giesberts, P., M.B.M. de Koster and J. Wijngaard, "Workload control in FMS environments", *paper presented at the DGOR Tagung*, September 1987.

Glassey, C.R. and Y. Hong, "The analysis of behavior of an unreliable two–stage automatic transfer line with inter–stage buffer storage", *Report ORC 86–5*, University of California, Berkeley, 1986a.

Glassey, C.R. and Y. Hong, "The analysis of behavior of an unreliable N–stage automatic transfer line with (N–1) inter–stage buffer storages", *Rep. ORC 86–10*, University of California, Berkeley, 1986b.

Gordon, W.J. and G.F. Newell, "Cyclic queueing systems with restricted queue lenghts", *Operations Research 15*, 266–277, 1967.

Gun, L. "Annotated bibliography of blocking systems", *Report SRC TR 87–187*, Systems Research Center, University of Maryland, 1987.

Gun, L. and A. Makowski, "Matrix–geometric solution for finite capacity queues with phase–type distributions", *Performance '87*, P.–J. Courtois and G. Latouche (eds.), 39–51, Elsevier, Amsterdam, 1988a.

Gun, L. and A.M. Makowski, "An approximation method for general tandem queueing systems subject to blocking", *presented at First International Workshop on Queueing Networks with Blocking*, Raleigh, 1988b.

Hadley, G. and T.M. Whitin, "Analysis of inventory systems", Prentice–Hall Inc., Englewood Cliffs, 1963.

Harrison, J.M., "The heavy traffic approximation for single server queues in series", *J. Applied Probability 10*, 613–629, 1973a.

Harrison, J.M., "Assembly–like queues", *J. Applied Probability 10*, 354–367, 1973b.

Harrison, J.M., "The diffusion approximation for tandem queues in heavy traffic", *Advances in Applied Probability 10*, 886–905, 1978.

Hatcher, J.M., "The effect of internal storage on the production rate of series of stages having exponential service times", *AIIE Transactions 1*, 150–156, 1969. (See also: Letter to the above article by A.D. Knott, *AIIE Transactions 2*, 273, 1970.

Hax, A.C. and D. Candea, *"Production and Inventory Management"*, Prentice Hall, 1984.

Hax, A.C. and H.C. Meal, "Hierarchical integration of production planning and scheduling", *Studies in Management Sciences, vol. I Logistics*, M.A. Geisler (ed.), North–Holland, Amsterdam, 1975.

Heidelberger, P. and S.K. Trivedi, "Queueing network models for parallel processing with asynchronous tasks", *IEEE Transactions on Computing 31*, 1099–1109, 1982.

Heidelberger, P. and S.K. Trivedi, "Analytic queueing models for programs with internal concurrency", *IEEE Transactions on Computing 32*, 73–82, 1983.

Hildebrand, D.K., "Stability of finite queue, tandem server systems", *J. Applied Probability 4*, 571–583, 1967.

Hildebrand, D.K., "On the capacity of tandem server, finite queue, service systems", *Operations Research 16*, 72–82, 1968.

Hillier, F.S. and R.W. Boling, "The effect of some design factors on the efficiency of production lines with variable operation times", *J. Industrial Eng. 17*, 651–658, 1966.

Hillier, F.S. and R.W. Boling, "Finite queues in series with exponential or erlang service times – A numerical approach", *Operations Research 15*, 286–303, 1967.

Ho, Y.C., Eyler, M.A. and T.T. Chien, "A gradient technique for general buffer storage design in a production line", *International J. Production Research 17*, 557–580, 1979.

Ho, Y.C., Eyler, M.A. and T.T. Chien, "A new approach to determine parameter insensitivities of transfer lines", *Management Science 29*, 700–714, 1983.

Hontelez, J. *"Modelling and Design of Production Lines* (in Dutch), Master's Thesis", Eindhoven University of Technology, 1985.

Hopp, W.J., N. Pati and P.C. Jones, "Optimal inventory control in a production flow system with failures", *Technical Report 88–07*, Northwestern University, Evanston, 1988.

Hunt, G.C., "Sequential arrays of waiting lines", *Operations Research 4*, 674–683, 1956.

Ignall, E. and A. Silver, "The output of a two–stage system with unreliable machines and limited storage", *AIIE Transactions 9*, 183–188, 1977.

Jun, K.P. and H.G. Perros, "Approximate analysis of arbitrary configurations of queueing networks with blocking and deadlock", *presented at First International Workshop on Queueing Networks with Blocking*,Raleigh, 1988.

Kay, E., "Buffer stocks in automatic lines", *Int. J. Prod. Res. 10*, 155– 165, 1972.

Kelly, F.P., "The throughput of a series of buffers", *Advances Applied Probability 14*, 633–653, 1982.

Kimemia, J. and S.B. Gershwin, "An algorithm for the computer control of a flexible manufacturing system", *IIE Transactions* 15, 353–362, 1983.

Klaren, M., *"An analyzing method for production lines with finite buffers and machine failures"*, Master's thesis (in Dutch), Twente University of Technology, 1987.

Kleinrock, L., *"Queueing Systems, vol. I: Theory"*, Wiley, New York, 1975.

Knott, A.D., "A letter to the editor", *AIIE Transactions 2*, 273, 1970.

Koenigsberg, E., "Production lines and internal storage – A review", *Management Science 5*, 410–433, 1959.

Kogan, Ya.A. and E.V. Krichagina, "Closed exponential queueing networks with blocking in heavy traffic", *presented at First International Workshop on Queueing Networks with Blocking*, Raleigh, 1988.

Kok, A.G. de, *"Production–Inventory Control Models; Approximations and Algorithms"*, CWI Tract 30, CWI, Amsterdam, 1987.

Konheim, A.G. and M. Reiser, "A queueing model with finite waiting room and blocking", *J. Association for Computing Machinery 23*, 328–341, 1976.

Konheim, A.G. and M. Reiser, "Finite capacity queueing systems with applications in computer modeling", *SIAM J. Computing 7*, 210–229, 1978.

Koster, M.B.M. de, "Approximation of flow lines with integrally controlled buffers", to appear in *IIE Transactions*, 1986a.

Koster, M.B.M. de, "Estimation of line efficiency by aggregation", *Report BDK/ORS/86/01*, Eindhoven University of Technology, 1986b.

Koster, M.B.M. de, "Capacity analysis of two–stage production lines with many products", *Engineering Costs and Production Economics 12*, 175–186, 1987a.

Koster, M.B.M. de, "Estimation of line efficiency by aggregation", *International J. Production Research 25*, 615–626, 1987b.

Koster, M.B.M. de, "Approximation of assembly–disassembly systems", *Report BDK/ORS/87/02*, Eindhoven University of Technology, 1987c.

Koster, M.B.M. de, "Approximate analysis of production systems", *European J. Operational Research 37*, 214–226, 1988a.

Koster, M.B.M. de, "An improved algorithm to approximate the behaviour of flow lines", *International J. Production Research 26*, 691–700, 1988b.

Koster, M.B.M. de and J. Wijngaard, "Approximations of two–stage production lines with intermediate buffer", *Report BDK/ORS/85/05*, Eindhoven University of Technology, 1985.

Koster, M.B.M. de and J. Wijngaard, "A continuous flow model for three production units in series with buffers", *Operations Research Proceedings 1985*, *DGOR*, Springer–Verlag Berlin, 253–264, 1986.

Koster, M.B.M. de and J. Wijngaard, "Local and integral control of workload", to appear in *International J. Production Research*, 1987a.

Koster, M.B.M. de and J. Wijngaard, "On the equivalence of multi–stage production lines and two–stage lines", *IIE Transactions 19*, 351–354, 1987b.

Koster, M.B.M. de and J. Wijngaard, "Capacity oriented logistic control", *Report BDK/ORS/88/01*, Eindhoven University of Technology, 1988a.

Koster, M.B.M. de and J. Wijngaard, "On the use of continuous vs discrete models in production lines with buffers", *presented at First International Workshop on Queueing Networks with Blocking*, Raleigh, 1988b.

Kouvatsos, D.D. and N.P. Xenios, "Maximum entropy analysis of general queueing networks with blocking", *presented at First International Workshop on Queueing Networks with Blocking*, Raleigh, 1988.

Kraemer, S.A. and R.G. Love, "A model for optimizing the buffer inventory storage size in a sequential production system", *AIIE Transactions 2*, 64–69, 1970.

Labetoulle, J. and G. Pujolle, "A study of queueing networks with deterministic service and applications to computer networks", *Acta Informatica 7*, 183–196, 1976.

Latouche, G. and M.F. Neuts, "Efficient algorithmic solutions to exponential tandem queues with blocking", *SIAM J. Algebraic Discrete Methods 1*, 93–106, 1980.

Latouche, G., "Queues with paired customers", *J. Applied Probability 18*, 1981

Lavenberg, S.S., "Stability and maximum departure rate of certain open queueing networks having finite capacity contraints", *RAIRO–Informatique/Comp.Science 12*, 353–370, 1978.

Lee, H.–S., and S.M. Pollock, "Approximate analysis of open exponential queueing networks with blocking: general configurations", *paper presented at First International Workshop on Queueing Networks with Blocking*, Raleigh, 1988.

Liu, X.–G. and J.A. Buzacott, "A balanced local flow technique for queueing networks with blocking", *presented at First International Workshop on Queueing Networks with*

Blocking, Raleigh, 1988.

Liu, Y.C. and H.G. Perros, "Approximate analysis of a closed fork/join model", *Center for Communication and Signal Processing report*, North Carolina State University, 1987.

MacGregor Smith, J. and S. Daskalaki, "Buffer space allocation in automated assembly lines", *Operations Research 36*, 343–358, 1988.

Magee, J.F., *"Production Planning and Inventory Control"*, McGraw–Hill, 1958.

Malathronas, J.P., Perkins, J.D. and R.L. Smith, "The availability of a system of two unreliable machines connected by an intermediate storage tank", *IIE Transactions 15*, 195–201, 1983.

Marie, R., "Calculating equilibrium probabilities for $\lambda(n)/C_k/1/N$ queues", *Performance 80*, 117–125, Toronto, 1980.

Masso, J. and M.L. Smith, "Interstage storages for three stage lines subject to stochastic failures", *AIIE Transactions 6*, 354–358, 1974.

Mattheij, R.M.M. and G.W.M. Staarink, "Boundpak user's manual", *report no. WD 85–03*, Catholic University of Nijmegen, 1985.

Meal, H.C., "Putting production decisions where they belong", *Harvard Business Review* 84, 102–111, 1984.

Melamed, B., "A note on the reversibility and duality of some tandem blocking queueing systems", *Management Science 32*, 1648–1650, 1986.

Meyer, R.R., Rothkopf, M.H. and S.A. Smith, "Reliability and inventory in a production–storage system", *Management Science 25*, 799–807, 1979.

Mitra, D., "Stochastic theory of a fluid model of multiple failure–susceptible producers and consumers coupled by a buffer", *to appear in Advances of Applied Probability*, 1986.

Mitra, D. "Stochastic fluid models", *Performance '87*, P.–J. Courtois and G. Latouche (eds.), 39–51, Elsevier, Amsterdam, 1988

Morse, P.M., *"Queues, Inventories and Maintenance"*, John Wiley & Sons, New York, 1965.

Murphy, R.A., "The effect of surge on system availability", *AIIE Trans. 7*, 439–443, 1975.

Murphy, R.A., "Estimating the output of a series production systems", *AIIE Transactions 10*, 139–148, 1978.

Murphy, R.A., "Examining the distribution of buffer protection", *AIIE Transactions 11*, 113, 1979.

Muth, E.J., "The production rate of a series of work stations with variable service times", *International J. Production Research 11*, 155–169, 1973.

Muth, E.J., "Numerical methods applicable to a production line with stochastic servers", *TIMS Studies in the Management Science 7*, 143–159, 1977.

Muth, E.J., "The reversibility property of production lines", Management Science 25, 152–158, 1979.

Muth, E.J. and A. Alkaff, "The throughput of three–station production lines: A unifying solution", *International J. Production Research 25*, 1405–1413, 1987.

NAG Fortran Library–Mark 12, Numerical Algorithms Group Ltd, Oxford, 1987.

Nelson, R. and N. Tantawi, "Approximate analysis of fork/join synchronization in parallel queues", *IBM Research Report* RC 11481, 1986.

Neuts, M.F., "Two queues in series with a finite intermediate waiting room", *J. Applied Probability. 5*, 123–142, 1968.

Neuts, M.F., "Two servers in series, studied in terms of a Markov renewal branching process", *Advances in Aplied Probability 2*, 110–149, 1970.

Neuts, M.F. *"Matrix–Geometric Solutions in Stochastic Models; An Algorithmic Approach"*, Johns Hopkins University Press, Baltimore, 1981

Newell, G.F., *"Approximate Behavior of Tandem Queues"*, Lecture Notes in Economics and Mathematicial Systems 171, Springer, Berlin, 1979.

Okamura, K. and H. Yamashina, "Analysis of the effect of buffer storage capacity in transfer line systems", *AIIE Transactions 9*, 127–135, 1977.

Onvural, R., "On the exact decomposition of closed queueing networks with finite buffers", *presented at First Int. Workshop on Queueing Networks with Blocking*, Raleigh, 1988.

Onvural, R.O. and H.G. Perros, "On equivalences of blocking mechanisms in queueing networks with blocking", *Operations Research Letters 5*, 293–297, 1986.

Patterson, R.L. "Markov processes occurring in the theory of traffic flow through and N–stage stochastic service system", *J. Industrial Engineering 15*, 188–193, 1964.

Paumen, P., *"Workload control. The foundation for production control of assembly factories of ICs at Philips"*, Master's Thesis, Eindhoven University of Technology, 1986.

Pease, M.C., "Methods of Matrix Algebra", *Academic Press*, New York, 1965.

Peck, L.G. and R.N. Hazelwood, *"Finite Queuing Tables"*, Wiley & Sons, New York, 1958.

Perros, H.G., "A survey of queueing networks with blocking, Part I", *Report Department of Computer Science*, North Carolina State University, Raleigh, 1986.

Perros, H.G. and T. Altiok, "Approximation analysis of open Networks of queues with blocking: Tandem configurations", *IEEE Trans. Software Eng. SE 12*, 450– 461, 1986.

Perros, H.G. and P.M. Snyder, "A computationally efficient algorithm for analyzing open queueing networks with blocking", *report NC State University*, 1986.

Plossl, G.W. and W.E. Welch, *"The Role of Top–Management in the Control of Inventory"*, Preston Pu. Co., 1979.

Pollock, S.M., Birge, J.R. and J.M. Alden, "Approximation analysis for open tandem queues with blocking; exponential and general service distributions", University of Michigan, Ann Arbor, 1986.

Prabhu, N.U., "Transient behavior of a tandem queue", *Management Science 13*, 631–639, 1967.

Pujolle, G., "The influence of protocols on the stability conditions in packet–switching networks", *to appear in IEEE Transactions on communications*, 1978.

Pujolle, G. and D. Potier, "Reseaux de file d'attente a capacite limitee avec des applications aux systemes informatiques", *RAIRO Informatique/Computer Science 13*, 175–197, 1979.

Ramsden, E., Ruane, M., Mavretic, A., Caramanis, M. and H. D'Angelo, "The use of hardware simulations in modeling production networks", *to appear in Large Scale Systems*, 1986.

Rao, N.P., "On the mean production rate of two–stage production system of the tandem type", *International J. Production Research 13*, 207–217, 1975a.

Rao, N.P., "Two–stage production systems with intermediate storage", *AIIE Transactions 7*, 414–421, 1975b.

Rao, N.P., "A viable alternative to the method of stages solution of Ssries production systems with Erlang service times", *International J. Production Research 14*, 699, 1976.

Reiser, M. and A.G. Konheim, "Finite capacity queueing systems with applications in computer modelling", *SIAM J. on Computing 7*, 210–229, 1978.

Ross, S.M., *"Applied Probability Models with Optimization Applications"*,Holden–Day, San Francisco, 1970.

Sauer, C.H. and K.M. Chandy, *"Computer Systems Performance Modelling"*, Prentice Hall, Englewood Cliffs, 1981.

Schick, I.C. and S.B. Gershwin, "Modelling and analysis of unreliable transfer lines with finite buffers", *MIT report ESL–FR–834–6*, 1978.

Schweitzer, P.J. and T. Altiok, "Aggregate modelling of tandem queues without intermediate buffers", *presented at First International Workshop on Queueing Networks with Blocking*, Raleigh, 1988.

Seneta, E., *Non–Negative Matrices. An Introduction to Theory and Applications*, George Allen & Unwin Ltd, London, 1973.

Sevast'yanov, B.A., "Influence of storage bin capacity on the average standstill time of a production line (in Russian)", *Teoriya Veroyatnostey i ee Primeniya 7*, 438–477 (Eng. tr. *Theory of Probability and Its Applications 7*, 429–438, 1962).

Shantikumar, J.G., "On the production capacity of automatic transfer lines with Unlimited buffer space", *AIIE Transactions 12*, 273, 1980.

Shantikumar, J.G. and Tien, "An algorithmic solution to two–stage transfer lines with possible scrapping of units", *Management Science 29*, 1069–1086, 1983.

Sheskin, T.J., "Allocation of interstage storage along an automatic production line", *AIIE Transactions 8*, 146–152, 1976.

Silver, E.A. and R. Peterson, *"Decision Systems for Inventory Management and Production Planning"*, Wiley Management and Administration Sciences (2nd ed.), 1985.

Soyster, A.L. Schmidt J.W. and M.W. Rohrer, "Allocation of buffer capacities for a class of fixed cycle production lines", *AIIE Transactions 11*, 140–146, 1979.

Soyster, A.L. and D.I. Toof, "Some comparative and design aspects of fixed cycle production systems", *Naval Research Logistics Quarterly 23*, 1976.

Suri, R. and G.W. Diehl, "A variable buffer–size model and its use in analyzing closed queueing networks with blocking", *Management Science 32*, 206–224, 1986.

Sutalaksana, I.Z., F.F. van Steelandt and L.N. van Wassenhove, "Integrating fixed scheduled maintenance and buffer storage–design for a two station transfer line",*Report 82–16*, Kath. Universiteit Leuven, 1982.

Sutalaksana, I.Z. and L.N. van Wassenhove, "Markov analysis of a two–station transfer line with buffer storage", *Report 81–24*, Kath. Universiteit Leuven, 1981.

Sutalaksana, I.Z. and L.N. van Wassenhove, "Markov analysis and simulation of a three station – two buffer line", *Report 82–08*, Kath. Universiteit Leuven, 1982.

Suzuki, T., "On a tandem queue with blocking", *J. Operations Research Society Japan 6*, 137–157, 1964.

Takahashi, Y., Miyahara, H. and T. Hasegawa, "An approximation method for open restricted queueing networks", *Operations Research 28*, 594–602, 1980.

Terracol, C. and R. David, "An aggregation method for performance evaluation of transfer lines with unreliable machines and finite buffers", *IEEE International conference on Robotics and Automation*, Raleigh, 1987.

Terracol, C. and R. David, "Performances d'une ligne composee de machines et de stocks intermediaires", *Automatique Productique Informatique Industrielle 21*, 239–262, 1987.

Timmer, J.P.J., J.W.M. Bertrand and W. Monhemius, "Production and inventory control with the Base Stock System", *Report BDK/KBS/84/05*, Eindhoven University of Technology, 1984.

Whitt, W., "Approximating a point process by a renewal process 1: two basic methods", *Operations Research 30*, 125–147, 1982.

Wiendahl, H.–P., *"Die belastungsorientierte Fertigungssteuerung – ein neues Steuerungsverfahren fur die Einzel– und Kleinserienfertigung"*, Institut fur Fabrikanlagen der Universitat Hannover, 1987

Wijngaard, J., "Aggregation and decomposition in logistic control of multi–stage, multi–product, production/inventory systems", *Operational Research '84*, J.P. Brans (ed.), 551–565, 1984.

Wijngaard, J., "Production control in a consumer electronics factory", *Engineering Costs and Production Economics 12*, 165–173, 1987.

Wijngaard, J., "The effect of interstage buffer storage on the output of two unreliable production units in series with different production rates", *AIIE Trans. 11*, 42–47, 1979.

Wong, B., Griffin, W. and R.L. Disney, "Two finite M/M/1 queues in tandem: a matrix solution for the steady state", *Opsearch 14*, 1–18, 1977.

Yamazaki, G., T. Kawashima and H. Sakasegawa, "Reversibility of tandem blocking queueing systems", *Management Science 31*, 78–83, 1985.

Yamazaki, G., H. Sakasegawa, T. Kawashima, "Production rate estimated with flow–shop reversibility", *Bulletin of the Japan Society of Mechanical Engineers 21*, 167–171, 1978.

Yeralan, S., Franck, W.E. and M.A. Quasem, "A continuous materials flow production line model with station breakdown", *European J. Operational Research 27*, 289–300, 1986.

Yeralan, S. and E.J. Muth, "A general model of a production line with intermediate buffer and station breakdown", *IIE Transactions 19*, 130–139, 1987.

Zimmern, B., "Etudes de la propagations des arrets aleatoires dans les chaines de production", *Revue de Statistiques Appliquees 4*, 85–104, 1956.

GLOSSARY OF NOTATIONS

notation	description	typical page reference
K, $\mathbf{K}$	buffer upper bound or capacity	29
λ, $\boldsymbol{\lambda}$	failure rate or life rate	29
ν, $\boldsymbol{\nu}$	machine speed	29
μ, $\boldsymbol{\mu}$	repair rate or life rate	29
M, $\mathbf{M}$	machine	29
B, $\mathbf{B}$	buffer	29
$v(K)$, $v(\mathbf{K})$	throughput of a system	31
E	expectation operator	31
X, $\mathbf{X}$	buffer content	31
C_T	expected cost, per cycle	31
$\alpha_i(x)$	cost rate in state (i,x)	31
$f(x)$	cost function till absorption with buffer level x	31
P	average production per cycle	33
PL	expected time, per cycle, that the buffer content is greater than, or equal to L	33
T	expected cycle time	33
I_ν	average time, per cycle, the line has input at rate ν	33
S_ν	average time, per cycle, the line has output at rate ν	33
O_i	average number of uptimes (downtimes) of the line input per cycle	33
O_0	average number of uptimes (downtimes) of the line output per cycle	33
Q_1	average time, per cycle, that the upstream machine is blocked	33
Q_2	average time per cycle, that the downstream machine is starved	33
Y	average buffer content	33
$\delta_{\{x\}}$	indicator. 1 if x holds, 0 if x does not hold	34
eff_i, $\mathbf{eff}$	efficiency or net average machine speed	36,53
σ	mean drift of a single buffer line	36,53

ρ	eigenvalue	37
P,Q,R	transition matrix of a machine	48
I(n)	the nxn unity matrix	49
O_I^z	average number of switches in a cycle, from an input speed smaller than or equal to z to a level greater than z	50
O_O^z	average number of switches in a cycle, from an output speed smaller than or equal to z to a level greater than z	50
$\mathbf{1}_n$	n–dimensional vector, consisting of ones	53
m	lower switching level	54
M	upper switching level	54
	set of machines in a network	113
k	buffer lower bound	55
$\beta_i(x)$	cost rate	57
U^i	increase of the unbounded buffer content of a two–stage line	63
U^d	decrease of the unbounded buffer content of a two–stage line	63
L^i	length of an increase period of the unbounded buffer content	63
L^d	length of a decrease period of the unbounded buffer content	63
cv	coefficient of variation	74
B	set of buffers in a network	113
$R_i(t)$	set of machines on a proliferating path with machine i at t	115
N	network of machines and buffers	116

General symbols:

$\mathbf{a}$,a_i	column vector (in Chapters 4,5,6,7 row vector)
A, a_{ij}	matrix
$\mathbf{a}^T$	transpose of a vector

INDEX

Vol. 236: G. Gandolfo, P.C. Padoan, A Disequilibrium Model of Real and Financial Accumulation in an Open Economy. VI, 172 pages. 1984.

Vol. 237: Misspecification Analysis. Proceedings, 1983. Edited by T.K. Dijkstra. V, 129 pages. 1984.

Vol. 238: W. Domschke, A. Drexl, Location and Layout Planning. IV, 134 pages. 1985.

Vol. 239: Microeconomic Models of Housing Markets. Edited by K. Stahl. VII, 197 pages. 1985.

Vol. 240: Contributions to Operations Research. Proceedings, 1984. Edited by K. Neumann and D. Pallaschke. V, 190 pages. 1985.

Vol. 241: U. Wittmann, Das Konzept rationaler Preiserwartungen. XI, 310 Seiten. 1985.

Vol. 242: Decision Making with Multiple Objectives. Proceedings, 1984. Edited by Y.Y. Haimes and V. Chankong. XI, 571 pages. 1985.

Vol. 243: Integer Programming and Related Areas. A Classified Bibliography 1981–1984. Edited by R. von Randow. XX, 386 pages. 1985.

Vol. 244: Advances in Equilibrium Theory. Proceedings, 1984. Edited by C.D. Aliprantis, O. Burkinshaw and N.J. Rothman. II, 235 pages. 1985.

Vol. 245: J.E.M. Wilhelm, Arbitrage Theory. VII, 114 pages. 1985.

Vol. 246: P.W. Otter, Dynamic Feature Space Modelling, Filtering and Self-Tuning Control of Stochastic Systems. XIV, 177 pages. 1985.

Vol. 247: Optimization and Discrete Choice in Urban Systems. Proceedings, 1983. Edited by B.G. Hutchinson, P. Nijkamp and M. Batty. VI, 371 pages. 1985.

Vol. 248: Plural Rationality and Interactive Decision Processes. Proceedings, 1984. Edited by M. Grauer, M. Thompson and A.P. Wierzbicki. VI, 354 pages. 1985.

Vol. 249: Spatial Price Equilibrium: Advances in Theory, Computation and Application. Proceedings, 1984. Edited by P. T. Harker. VII, 277 pages. 1985.

Vol. 250: M. Roubens, Ph. Vincke, Preference Modelling. VIII, 94 pages. 1985.

Vol. 251: Input-Output Modeling. Proceedings, 1984. Edited by A. Smyshlyaev. VI, 261 pages. 1985.

Vol. 252: A. Birolini, On the Use of Stochastic Processes in Modeling Reliability Problems. VI, 105 pages. 1985.

Vol. 253: C. Withagen, Economic Theory and International Trade in Natural Exhaustible Resources. VI, 172 pages. 1985.

Vol. 254: S. Müller, Arbitrage Pricing of Contingent Claims. VIII, 151 pages. 1985.

Vol. 255: Nondifferentiable Optimization: Motivations and Applications. Proceedings, 1984. Edited by V.F. Demyanov and D. Pallaschke. VI, 350 pages. 1985.

Vol. 256: Convexity and Duality in Optimization. Proceedings, 1984. Edited by J. Ponstein. V, 142 pages. 1985.

Vol. 257: Dynamics of Macrosystems. Proceedings, 1984. Edited by J.-P. Aubin, D. Saari and K. Sigmund. VI, 280 pages. 1985.

Vol. 258: H. Funke, Eine allgemeine Theorie der Polypol- und Oligopolpreisbildung. III, 237 pages. 1985.

Vol. 259: Infinite Programming. Proceedings, 1984. Edited by E.J. Anderson and A.B. Philpott. XIV, 244 pages. 1985.

Vol. 260: H.-J. Kruse, Degeneracy Graphs and the Neighbourhood Problem. VIII, 128 pages. 1986.

Vol. 261: Th.R. Gulledge, Jr., N.K. Womer, The Economics of Made-to-Order Production. VI, 134 pages. 1986.

Vol. 262: H.U. Buhl, A Neo-Classical Theory of Distribution and Wealth. V, 146 pages. 1986.

Vol. 263: M. Schäfer, Resource Extraction and Market Structure. XI, 154 pages. 1986.

Vol. 264: Models of Economic Dynamics. Proceedings, 1983. Edited by H.F. Sonnenschein. VII, 212 pages. 1986.

Vol. 265: Dynamic Games and Applications in Economics. Edited by T. Başar. IX, 288 pages. 1986.

Vol. 266: Multi-Stage Production Planning and Inventory Control. Edited by S. Axsäter, Ch. Schneeweiss and E. Silver. V, 264 pages. 1986.

Vol. 267: R. Bemelmans, The Capacity Aspect of Inventories. IX, 165 pages. 1986.

Vol. 268: V. Firchau, Information Evaluation in Capital Markets. VII, 103 pages. 1986.

Vol. 269: A. Borglin, H. Keiding, Optimality in Infinite Horizon Economies. VI, 180 pages. 1986.

Vol. 270: Technological Change, Employment and Spatial Dynamics. Proceedings 1985. Edited by P. Nijkamp. VII, 466 pages. 1986.

Vol. 271: C. Hildreth, The Cowles Commission in Chicago, 1939–1955. V, 176 pages. 1986.

Vol. 272: G. Clemenz, Credit Markets with Asymmetric Information. VIII, 212 pages. 1986.

Vol. 273: Large-Scale Modelling and Interactive Decision Analysis. Proceedings, 1985. Edited by G. Fandel, M. Grauer, A. Kurzhanski and A.P. Wierzbicki. VII, 363 pages. 1986.

Vol. 274: W.K. Klein Haneveld, Duality in Stochastic Linear and Dynamic Programming. VII, 295 pages. 1986.

Vol. 275: Competition, Instability, and Nonlinear Cycles. Proceedings, 1985. Edited by W. Semmler. XII, 340 pages. 1986.

Vol. 276: M.R. Baye, D.A. Black, Consumer Behavior, Cost of Living Measures, and the Income Tax. VII, 119 pages. 1986.

Vol. 277: Studies in Austrian Capital Theory, Investment and Time. Edited by M. Faber. VI, 317 pages. 1986.

Vol. 278: W.E. Diewert, The Measurement of the Economic Benefits of Infrastructure Services. V, 202 pages. 1986.

Vol. 279: H.-J. Büttler, G. Frei and B. Schips, Estimation of Disequilibrium Models. VI, 114 pages. 1986.

Vol. 280: H.T. Lau, Combinatorial Heuristic Algorithms with FORTRAN. VII, 126 pages. 1986.

Vol. 281: Ch.-L. Hwang, M.-J. Lin, Group Decision Making under Multiple Criteria. XI, 400 pages. 1987.

Vol. 282: K. Schittkowski, More Test Examples for Nonlinear Programming Codes. V, 261 pages. 1987.

Vol. 283: G. Gabisch, H.-W. Lorenz, Business Cycle Theory. VII, 229 pages. 1987.

Vol. 284: H. Lütkepohl, Forecasting Aggregated Vector ARMA Processes. X, 323 pages. 1987.

Vol. 285: Toward Interactive and Intelligent Decision Support Systems. Volume 1. Proceedings, 1986. Edited by Y. Sawaragi, K. Inoue and H. Nakayama. XII, 445 pages. 1987.

Vol. 286: Toward Interactive and Intelligent Decision Support Systems. Volume 2. Proceedings, 1986. Edited by Y. Sawaragi, K. Inoue and H. Nakayama. XII, 450 pages. 1987.

Vol. 287: Dynamical Systems. Proceedings, 1985. Edited by A.B. Kurzhanski and K. Sigmund. VI, 215 pages. 1987.

Vol. 288: G.D. Rudebusch, The Estimation of Macroeconomic Disequilibrium Models with Regime Classification Information. VII, 128 pages. 1987.

Vol. 289: B.R. Meijboom, Planning in Decentralized Firms. X, 168 pages. 1987.

Vol. 290: D.A. Carlson, A. Haurie, Infinite Horizon Optimal Control. XI, 254 pages. 1987.

Vol. 291: N. Takahashi, Design of Adaptive Organizations. VI, 140 pages. 1987.

Vol. 292: I. Tchijov, L. Tomaszewicz (Eds.), Input-Output Modeling. Proceedings, 1985. VI, 195 pages. 1987.

Vol. 293: D. Batten, J. Casti, B. Johansson (Eds.), Economic Evolution and Structural Adjustment. Proceedings, 1985. VI, 382 pages. 1987.

Vol. 294: J. Jahn, W. Krabs (Eds.), Recent Advances and Historical Development of Vector Optimization. VII, 405 pages. 1987.

Vol. 295: H. Meister, The Purification Problem for Constrained Games with Incomplete Information. X, 127 pages. 1987.

Vol. 296: A. Börsch-Supan, Econometric Analysis of Discrete Choice. VIII, 211 pages. 1987.

Vol. 297: V. Fedorov, H. Läuter (Eds.), Model-Oriented Data Analysis. Proceedings, 1987. VI, 239 pages. 1988.

Vol. 298: S.H. Chew, Q. Zheng, Integral Global Optimization. VII, 179 pages. 1988.

Vol. 299: K. Marti, Descent Directions and Efficient Solutions in Discretely Distributed Stochastic Programs. XIV, 178 pages. 1988.

Vol. 300: U. Derigs, Programming in Networks and Graphs. XI, 315 pages. 1988.

Vol. 301: J. Kacprzyk, M. Roubens (Eds.), Non-Conventional Preference Relations in Decision Making. VII, 155 pages. 1988.

Vol. 302: H.A. Eiselt, G. Pederzoli (Eds.), Advances in Optimization and Control. Proceedings, 1986. VIII, 372 pages. 1988.

Vol. 303: F.X. Diebold, Empirical Modeling of Exchange Rate Dynamics. VII, 143 pages. 1988.

Vol. 304: A. Kurzhanski, K. Neumann, D. Pallaschke (Eds.), Optimization, Parallel Processing and Applications. Proceedings, 1987. VI, 292 pages. 1988.

Vol. 305: G.-J.C.Th. van Schijndel, Dynamic Firm and Investor Behaviour under Progressive Personal Taxation. X, 215 pages. 1988.

Vol. 306: Ch. Klein, A Static Microeconomic Model of Pure Competition. VIII, 139 pages. 1988.

Vol. 307: T.K. Dijkstra (Ed.), On Model Uncertainty and its Statistical Implications. VII, 138 pages. 1988.

Vol. 308: J.R. Daduna, A. Wren (Eds.), Computer-Aided Transit Scheduling. VIII, 339 pages. 1988.

Vol. 309: G. Ricci, K. Velupillai (Eds.), Growth Cycles and Multisectoral Economics: the Goodwin Tradition. III, 126 pages. 1988.

Vol. 310: J. Kacprzyk, M. Fedrizzi (Eds.), Combining Fuzzy Imprecision with Probabilistic Uncertainty in Decision Making. IX, 399 pages. 1988.

Vol. 311: R. Färe, Fundamentals of Production Theory. IX, 163 pages. 1988.

Vol. 312: J. Krishnakumar, Estimation of Simultaneous Equation Models with Error Components Structure. X, 357 pages. 1988.

Vol. 313: W. Jammernegg, Sequential Binary Investment Decisions. VI, 156 pages. 1988.

Vol. 314: R. Tietz, W. Albers, R. Selten (Eds.), Bounded Rational Behavior in Experimental Games and Markets. VI, 368 pages. 1988.

Vol. 315: I. Orishimo, G.J.D. Hewings, P. Nijkamp (Eds.), Information Technology: Social and Spatial Perspectives. Proceedings, 1986. VI, 268 pages. 1988.

Vol. 316: R.L. Basmann, D.J. Slottje, K. Hayes, J.D. Johnson, D.J. Molina, The Generalized Fechner-Thurstone Direct Utility Function and Some of its Uses. VIII, 159 pages. 1988.

Vol. 317: L. Bianco, A. La Bella (Eds.), Freight Transport Planning and Logistics. Proceedings, 1987. X, 568 pages. 1988.

Vol. 318: T. Doup, Simplicial Algorithms on the Simplotope. VIII, 262 pages. 1988.

Vol. 319: D.T. Luc, Theory of Vector Optimization. VIII, 173 pages. 1989.

Vol. 320: D. van der Wijst, Financial Structure in Small Business. VII, 181 pages. 1989.

Vol. 321: M. Di Matteo, R.M. Goodwin, A. Vercelli (Eds.), Technological and Social Factors in Long Term Fluctuations. Proceedings. IX, 442 pages. 1989.

Vol. 322: T. Kollintzas (Ed.), The Rational Expectations Equilibrium Inventory. XI, 269 pages. 1989.

Vol. 323: M.B.M. de Koster, Capacity Oriented Analysis and Design of Production Systems. XII, 245 pages. 1989.